"For years, Les Gelb's friends have been learning about foreign policy by way of his wisecracks and anecdotes. In *Power Rules*, he shares a lifetime's worth of wit and wisdom with the rest of the class. The amazing thing about this shrewd updating of *The Prince* is not just the insight Gelb brings to the topic of America's exercise of power in the post–Cold War, post-Bush world, but how entertaining he makes the whole subject. This book is a must-read not just for President Obama, but for anyone who wants to understand how the new administration can improve its odds of strategic success."
                                    —Jacob Weisberg

"A thoughtful, comprehensive, and engaging examination. . . . Gelb's bulleted rules and clear advice to President Obama distill his moderate strategic thinking on the future of America. It is a vision of a pragmatic but responsible global U.S. presence that eschews partisan politics and should find favor in the coming political clime."
            —*Publishers Weekly* (starred review/Pick of the Week)

"Les Gelb tells it like it is: making U.S. foreign policy and using American power are common sense, not rocket science. Our leaders forget this truth at our peril. Incisive and thoroughly compelling, *Power Rules* is rich in colorful stories as well as in sound advice for our president and our people."   —Brent Scowcroft

"It is to Gelb's credit that *Power Rules* is striking more for its similarities to than its differences with *The Prince*. Like Machiavelli, Gelb gets to the nub of hard truths about power and statecraft in a complex world. . . . Compared to the piles of books being churned out about America's place in the world, *Power Rules* belongs in the top tier. Gelb intended this book as a long letter to President Obama; I fervently hope that the intended recipient reads it carefully."   —*The National Interest*

POWER
RULES

POWER RULES

HOW COMMON SENSE CAN RESCUE
AMERICAN FOREIGN POLICY

Leslie H. Gelb

HARPER PERENNIAL

NEW YORK • LONDON • TORONTO • SYDNEY • NEW DELHI • AUCKLAND

JUDY, JUDY, JUDY

HARPER ● PERENNIAL

A hardcover edition of this book was published in 2009 by HarperCollins
Publishers.

HarperCollins books may be purchased for educational, business, or sales
promotional use. For information, please e-mail the Special Markets Department at SPsales@harpercollins.com.

FIRST HARPER PERENNIAL EDITION PUBLISHED 2010.

Designed by Renato Stanisic

The Library of Congress has catalogued the hardcover edition as follows:
Gelb, Leslie H.
Power rules: how common sense can rescue American foreign policy / Leslie H.
Gelb.—1st ed.   p. cm.
Includes bibliographical references and index.
ISBN 978-0-06-171454-2
1. United States—Foreign relations—2001.   2. Power (Social sciences)—United
States.   3. World politics—21st century.   I. Title.
JZ1480.A5G45 2009
327.73—dc22   2008051977

ISBN 978-0-06-171456-6 (pbk.)

20   ov/LSC   10 9 8 7 6 5 4

Contents

Contents

LETTER TO OUR ELECTED PRINCE

When Niccolò Machiavelli addressed his book about power—the favorite subject of princes, premiers, professors, and presidents—to Prince Lorenzo de' Medici in 1513, he implored him to "gaze down from the summit of your lofty position toward this humble spot, [where] you will recognise the great and unmerited sufferings inflicted on me by a cruel fate." The author of *The Prince* was pleading with Lorenzo to restore his lost government post. His masterpiece, in other words, was a job application.

This book is also about power, but it is not a job application. It may even fulfill my quest for the model anti–government job application. For, as I call your attention to the missteps on the road to gaining and retaining power, as Machiavelli did for his prince nearly five centuries ago, I am sure to offend many of the missteppers, just as the power master did. Officials (and would-be officials) still haven't learned to appreciate even the most heartfelt and generous public advice on how to repair their mistakes.

Nor is this a partisan book. I am a frequent and equal-opportunity critic. I have also served both Republicans and Democrats. In recent years, however, I have become much more a critic than an admirer of both parties. That applies especially to George W. Bush, but also to William Jefferson Clinton. Both men and their parties have made me not partisan, but just a bit surly.

Nor is this a book on how to transform the world into a democratic, threat-free, free-market paradise. I have long believed with Machiavelli that visionaries do more harm than good, and have therefore subdued my enthusiasm for them with this saying: Without vision, men die; with vision, more men die. My hope is to offer a clear sense of direction, but not a vision.

This book, I must admit to you, is more about reminders than revelations. We need to be reminded of the proven ways to think sensibly about power and policy. As you feed on these reminders of old and forgotten wisdom, I hope you will stumble upon some useful revelations as well—revelations on what's old and new about power in the world of the twenty-first century, and how to use today's novel blends of power.

My career has spanned three worlds that center on power: government, journalism, and think tanks. I was executive assistant to Jacob Javits, the Republican senator from New York; then later, director of policy planning and arms control in the Pentagon under President Lyndon B. Johnson; and later still, assistant secretary for politico-military affairs in the State Department under President Jimmy Carter. In my years with *The New York Times*, I worked as national security correspondent, op-ed page editor, and foreign affairs columnist. And now I've become a septuagenarian blogger for *The Daily Beast*. Intermittently, I was a senior fellow at the Brookings Institution and the Carnegie Endowment for International Peace, both in Washington, D.C., and served as president of the Council on Foreign Relations in New York City.

My interest in international affairs and power began when I earned my doctorate at Harvard under the direction of professors Stanley Hoffmann and Henry Kissinger. The latter's most memorable advice to me still resonates: "To be profound, it is necessary to be obscure." I fear that in this book I shall once again disappoint him.

I write this book for the millions of Americans and others around the world who have been bewildered and embittered for decades about the endless parade of grievous and terribly costly U.S. foreign policy mistakes, and about the lack of common sense in our foreign policy. Intuitively, most Americans sense that there must be a better way—and they are right. This book shows them why they are right.

But I address this book to you, President Obama (and to your successors, of course), just as Machiavelli wrote for his prince, because when it comes to foreign policy—even in our twenty-first-century democracy—presidents are princes. You make policy, and you decide on war or peace.

By contrast with princes of yore, however, you are an elected prince. That gives my fellow citizens the responsibility to learn enough about the world and policy alternatives to judge you.

While you, Mr. President, are the main addressee for this book, power is its star. Permit me a final admonition, then, that brings you and power together before you embark on these pages: Power is power. It is neither hard nor soft nor smart nor dumb. Only you can be hard, soft, smart, or dumb.

INTRODUCTION

I t was reserved for Augustus to relinquish the ambitious design of subduing the whole earth, and to introduce a spirit of moderation into the public councils," wrote Edward Gibbon in *The History of the Decline and Fall of the Roman Empire*, first published in 1776. "Inclined to peace by his temper and situation, it was easy for him to discover that Rome, in her present exalted situation, had much less to hope than to fear from the chance of arms; and that, in the prosecution of remote wars, the undertaking became every day more difficult, the event more doubtful, and the possession more precarious, and lest [*sic*] beneficial. The experience of Augustus added weight to these salutary reflections, and effectually convinced him that, by the prudent vigour of his counsels, it would be easy to secure every concession which the safety or the dignity of Rome might require from the most formidable Barbarians."

I quote Gibbon here not to suggest the decline and fall of the United States, although that thought can't be dismissed in these troubled times. Rather, I quote him to note his judgment that after years of turmoil and imperial overreach, Emperor Caesar Augustus put a sensible understanding of power back into Roman power.

My first goal in this book is to put power back into American power, to fit it to twenty-first century realities and thus make it effective again. My second is to restore common sense to the exercise of that power and the making of American foreign policy.

Power is as vital today as ever in securing national interests. It remains the necessary means to all important international ends, the principal coin of the global realm.

Power rules, still, and there still are rules on how best to exercise it. The rules differ from those penned by Niccolò Machiavelli in *The Prince* almost 500 years ago, but share the same roots in a mixed view of human nature and a constant sense of the uncertainties of a semi-anarchic world.

The problem is that the core meaning of power has been lost, or even worse, hijacked by various liberals and conservatives in a constant and all-consuming battle. These warriors chose their battleground well. They knew that whoever defines power controls U.S. foreign policy. As they contended, power became more an ideological weapon in internal political wars than an instrument of foreign policy.

The first task now is to clear away the smoke and take back the discussion of power from the looters and the fashion designers of international policy, whose creations have temporarily delighted those always searching for new and big truths. That means jousting with the leading voices of our time—the soft and hard powerites, America's premature grave diggers, the world-is-flat globalization crowd, and the usually triumphant schemers who ceaselessly demand that America "must do" certain things regardless of their achievability.

Power is not soft, or hard, for that matter. It is what it always was—essentially the capacity to get people to do what they don't want to do, by pressure and coercion, using one's resources and position. The idea is to cause others to worry about what you can do for them or to them. Persuasion, values, and the use of force can and often do flow into power, but at its core, power is psychological and political pressure.

We are not in, nor are we entering, a post-American era. In the first place, this fashionable observation incorrectly suggests that the United States dominated the Cold War era. But Washington was not the master of that bipolar universe; the Soviet Union was also a superpower. In the second place, Washington did not even rise to the role of shaper, let alone dictator, in the fleeting and hopeful window between the collapse of the Soviet Union and today. The much-predicted post-American era may indeed be on the horizon, if our economy and government falter, and if we continue to butcher our foreign policy. But we're not there yet.

The world is not flat—that is, flattened by economic globalization, the information revolution, and the equalization of power. The shape of global power is decidedly pyramidal—with the United States alone at the top, a second tier of major countries (China, Japan, India, Russia, the United Kingdom, France, Germany, and Brazil), and several tiers descending below. Even the smallest countries now occupy a piece of the international pyramid and have, particularly, enough power to resist the strong. But among all nations, only the United States is a true global power with global reach.

To be at the top, even alone, does not give the United States the power to dominate; there is too much resisting power now sloshing around in the world for that. But being the sole occupant of the world pyramid's penthouse does provide one absolutely critical attribute: It leaves the United States with the unique power to lead, not anywhere it wants, but toward the solution of major international problems such as issues of trade, security, and the environment. Most countries know full well that if Washington fails to organize action on a major issue, nothing is likely to get done. Thus, the power to lead derives from the power to solve problems in the interests of the key nations involved.

Nor, as some would argue, is the world nonpolar in the sense that there are no important concentrations of power. The world is a blend of unipolarity and multipolarity. The United States is the indispensable leader, the only nation capable of leading with regard to the world's key threats and opportunities. No one can seriously doubt this kind of unipolarity. It makes the United States the paramount power, but not by any stretch the dictator in war or peace.

This, in turn, means that Washington can't solve major problems on its own. Unilateral action even in military extremis isn't likely to work. Washington couldn't even begin to fight in Afghanistan, for example, without dozens of countries allowing overflights and bases providing many other forms of cooperation. This is where multipolarity comes in: Other key nations—which also can't solve problems, either on their own or together—are equally indispensable as partners in solving problems.

This clear mutual dependence makes mutual indispensability the central operating principle for power in the twenty-first century. Beyond reasonable argument, it means this: We fail alone, but can succeed together. We need others and we have the requisite power to lead them to solve problems, and they most assuredly need us. Recognition of mutual indispensability, alas, does not guarantee common action, given the separatist tugs of internal politics and perceived national interests. It does, however, point leaders in the right direction.

This concept is most certainly not a recipe for a policy of weak-kneed multilateralism or the orgiastic massing of nations committed to inaction on their own or through the United Nations. Rather, it prompts Washington to forge power coalitions of key and relevant states whose combined powers can solve the problem at hand. This is not a disguise for old-fashioned multilateralism or a sheep in wolf's clothing; it is a commonsensical formula for exercising power successfully in the twenty-first century.

There is a final and essential battle to win in order to restore the *power* of American power: It is to defang those liberals and conservatives who repeatedly corner our leaders into making commitments they cannot fulfill. America has endured more than half a century's worth of these unattainable goals. We live with them now: nation-building in places such as Afghanistan, where there is no coherent nation and certainly no outsider could do this for them anyway; spreading democracy to countries such as Iraq with no tradition of, or foundations for, democracy; and insisting on bringing such places as Georgia or Ukraine under NATO's wing with neither the intent nor the capability of actually sending troops to defend them. False promises and failures are the surest way to kill power.

None of this is to argue that there are not situations and mo-

ments when normal prudence must give way to national commitment, national willpower, and national sacrifice. This is the only
possible response to the Hitlers and present-day terrorists who must
be convinced of our unrelenting resolve and power to defeat them.
But this degree of resolve and sacrifice should be reserved for only
the most palpable and dire dangers.

To me, all these analyses, judgments, and prescriptions flow from
common sense. To me, all power and policy issues are best governed
by common sense, and not elevated into rocket science. But our
leaders forever find themselves being dragged off practical paths by
the demons of principles, politics, and the arrogance of power—and
into tragic failures.

In the end, this is a book that tries to restore the effectiveness of
American power by fighting for good old American common sense.
Augustus was never so foolish as to skimp on the Roman legions or
neglect Rome's economy. But on the basis of those assets and Rome's
formidable position in its world, he came to employ "the prudent
vigour of his counsels"—meaning his tough yet restrained pressure
on others—as the principal means to exercise power and protect
Roman interests even "from the most formidable Barbarians."

Part I

Power in the New World

CHAPTER 1

THE REVOLUTION IN WORLD POWER

Here's the central paradox of twenty-first-century world affairs: The United States is probably the most powerful nation in history, yet far more often than not, it can't get its way. The 500-year story that led up to the current state of affairs reveals the new and revolutionary rules and rhythms of international power.

Fidel Castro's Cuba, one of the world's smaller and weaker nations, gave constant strategic and political grief to the United States, the world's strongest, and survived to tell the tale for almost half a century. In any previous era, a major power like the United States would have quickly and violently crushed such a pesky little neighbor. But America's forbearance toward Cuba epitomizes a profound and underappreciated story, the story of the reshaping and rechanneling of international power—nothing short of a revolution in the history of world affairs.

Leaving aside philosophical rights and wrongs, the fact is that Castro flouted all the rules of deference toward great powers, especially one only ninety miles away. He joined with the Soviet Union, America's mortal foe, in precipitating the 1962 missile crisis, the single most dangerous moment of the Cold War. Later, he picked at the sores by sending his troops to fight the United States' allies in Africa and Latin America.

By historical standards, Washington had more than sufficient grounds to overthrow Castro. Instead, one by one, U.S. presidents, including some of the toughest, restricted themselves to feeble and futile efforts to spark anti-Castro revolutions (e.g., the ill-fated Bay of Pigs invasion of 1961) and strangle him economically. No American president seriously undertook the only action guaranteed to rid Washington of Castro—an all-out invasion of Cuba.

To be sure, the clever Cuban dictator helped stave off attacks by perching under the Soviet wing, waiting for Moscow and Washington to agree to live and let live in Cuba. But such distant protection and promises of restraint from great powers had rarely held up in the past. Historically, protectors like the Soviet Union readily sold out a client like Cuba for a better power deal somewhere else.

Castro surely sensed other, deeper restraints on America's power. He clearly understood that U.S. presidents, for all their bellicose Cold War rhetoric, dared not invade Cuba without clear provocation. He must have grasped that the restraints of world opinion and American democracy, as well as the prospect of fighting a determined insurgency inside Cuba, had come to matter a great deal and would stay the hand of even the fiercest U.S. president.

Castro's Cuba was both the beneficiary and the symbol of profound changes in the rules and rhythms of international power. The seeds of this revolution ran deep into history, took hold during the Cold War, and then fully rooted in the twenty-first century. During those almost five decades, the number of nation-states multiplied, most of them with the political will and new means to resist domination by the great powers. Worldwide communications expanded exponentially with the effect of informing and exciting peoples against the great powers' machinations. International commerce took unprecedented flight, creating common interests and restraints on rich and poor nations alike. And nuclear weapons fundamentally altered the role of military force in traditional big power rivalries.

It was hard to get a solid fix on how power was changing, and on what was old and what was new in international affairs. But three patterns began to emerge.

First, the strong fled from direct military confrontations with one another, instead of following their time-honored pattern of resolving their differences by war. Nuclear weapons especially made big power military contests too destructive and dangerous. Traditional conventional war had become much too expensive as well, especially when governments assumed greater responsibility for their citizens' welfare. Even hawks found it increasingly difficult to define national interests in such a way as to justify risking the catastrophe of nuclear war.

Second, the power of the weak to resist the strong started to rival the power of the strong to command—at least on the weak's own turf. Backed by a slew of new international constraints on the strong, the weak frequently challenged the strong—and often got away with it.

Third, while traditional balance-of-power competition continued to mark twenty-first-century international affairs, competition over vital interests was not as ferocious as before. Big and small states alike increasingly turned to a vast and relatively new array of international institutions and norms to protect their interests. When these proved ineffective or required supplementing, most nations resorted to the old balance-of-power reflexes.

All these twenty-first-century patterns of power are underpinned and reinforced by two earthquaking historical trends: the declining utility of military power and the concomitant rise of international economic power. Military capability—both the threat and the use of force—still counts significantly, but today, as compared with the past, there are more uncertainties about its use. At the same time, economic strength has increased in importance, both as an instrument of international power and as a restraint on it. This is a mysterious form of power, far more complex to wield than sheer military force, the mother of all blunt instruments.

The net effect of the new patterns of power and the underlying changes in military and economic power do not negate the importance of power, but they do restrain and complicate its use. Power continues to matter more than anything else in international transactions. Ideas, leadership, and appeals to reason can mobilize peoples to revolt against tyrants and persuade citizens to make sacrifices within nations; but they rarely lead to changes in government policies, and they have a poor track record in resolving conflicts between and among nations. For this, economic and other benefits bestowed and withheld, the twin instruments of pressure and coercion, still prove the better, if not the only, means of getting things done internationally. But this crucial instrument—power—cannot be used effectively in the twenty-first century without an understanding of how the new constraints on it have evolved.

THE MAIN STORY LINE of the last 500 years—that dominant nations can no longer dominate weaker ones at will—seemed unimaginable at the dawn of the modern era in the sixteenth century, when kings and princes created early versions of what is now known as the nation-state. Their conquests would gather vast power around the world, even as they sowed the seeds that would lead to the loss of their far-flung colonies.

The dawn of the modern state came at the time of Niccolò Machiavelli, the great power master. His time, nearly half a millennium ago, was far different from today. A handful of nation-states in Europe were just beginning to take their present geographical and political shape, and they often turned to military force as the arbiter of all international disputes, large or small. As these new nation-states gained superiority in military power, they became the new stars in the European firmament and, soon, in the world as well.

Machiavelli's Italy, however, was neither a nation nor a state. It was a peninsula dotted with city-states, which were actually just cities. Despite their small size, their names linger to this day for their great artistic accomplishments, their crimes, and their follies: Rome, Venice, Milan, and Machiavelli's beloved Florence among them. These hyperactive mini-entities conspired and warred against one another until, inevitably, the weaker ones reached far afield for help from the new giant nation-states of Europe. To Spain, France, and the Holy Roman Empire, the envoys of the city-states journeyed to seek aid.

The princes of Italy strategically positioned themselves under the wings of European kings and their mercenary armies. Kings dispatched their armies to do battle, exacting a heavy price for the protection they afforded. This pitiless European game came to be known as the balance-of-power system. The idea was to gather superior military force through alliances or to ward off impending imbalances through counter-alliances.

The theory is simple to explain but was difficult to operate. According to the great historian of Renaissance diplomacy, Garrett Mattingly, the Italians believed they could avoid disaster because they thought they were smarter than everyone else. Many, Mattingly argued, reckoned they could compensate for their lack of military prowess by means of their superior intellect and their command of the arts of negotiation. To their northern neighbors and to Machiavelli as well, the truth was perhaps best expressed by a dictum of Prussia's King Frederick the Great: "Diplomacy without arms is like a concert without a score."

Mattingly described it as a war of all against all: "Shiftiness and inconstancy were imposed on the Italian system by the internal political instability of most of the major states, by the delicate balance of peninsular power, and, chiefly, by the continuous struggle of each state against all." Mattingly wrote unsparingly of the Renaissance emperors, kings, and princes as they developed the rules of their costly and tragic games. "Nor in the decade in which by invading Italy she began the age of modern European diplomacy had France any coherent foreign policy, either. She went to war simply because it was always assumed that when Charles VIII came of age he would go to war. What else could a young, healthy king with money in his treasury and men-at-arms to follow him be expected to do? War was the business of kings."

From about 1500 to the end of World War I in 1918, the rulers of Europe and their ministers devoted far more time and resources to playing balance-of-power games and waging wars than to caring for their subjects. Nothing, it seemed, was more fun for them than conquering a neighbor's territory, population, and resources, thereby adding to their own riches and power. Invariably, their wars generated more grievances that led to new wars. When warring on the European chessboard proved too costly, inconclusive, or boring, the kings created a new chessboard composed of their colonial territories, sending their armies across oceans and continents to conquer peoples in faraway lands, not only to expand beyond their neighbors' borders but also to build even larger empires.

International power was the ultimate expression of power, and military prowess represented the ultimate expression of international power. Machiavelli saw this simple and brutal fact with total clarity, as did four subsequent centuries of European leaders; that is why his words continued to resonate with them. Machiavelli advised rulers to prepare for and make war, deliver a hard and smart peace, and then prepare for war again. And for 450 years, that's exactly what they did. Commerce both inspired and followed the flag, but conquest was the name of the game in the great age of empires—conquest pursued as much for its own sake as for the economic gains it brought.

Napoleon took this game to new destructive heights by perfecting the nation-state. Until he applied his genius to the task, most nation-states were poorly organized behemoths. Napoleon combined the nation and the state—arms, men, resources, and a type of nationalism—into a fighting force unmatched by any other single power of his time. And because he was the first to do this, he conquered most of Europe—until Europe finally united to defeat him.

As Napoleon was reaching the zenith of his power, his conquests began to produce ripple effects not only in Europe, but in the Americas as well. The fervor of nationalism unleashed by the French Revolution helped propel Napoleon's plan for transforming France into the powerhouse of Europe. But this same fervor also proved his undoing. Other nations discovered their own sense of nationhood and used it to fight for their independence against Napoleon. The Spanish successfully resisted him in the peninsular campaigns of 1808. The Russians outlasted him, despite Napoleon's capture of Moscow. Nationalism was beginning to arrive in Europe, never to depart.

Similar forces had appeared in the Americas. Before the nineteenth century, the American and Haitian revolutions foreshadowed distant and future global trends. In the name of independence, George Washington's ragtag army (with eventual French help, to be sure) defeated the largest expeditionary force that the United Kingdom, the world's greatest imperial power, had ever dispatched abroad. Some years later, in the name of freedom, Haitian slaves thwarted another huge British army and negotiated its withdrawal, and later, thwarted mighty France as well. These defeats of the strong by the weak—indeed, of the strongest by the weakest—foretold profound shifts in power that would remain quietly submerged for another century and a half before surfacing. Forces larger and more profound than power based on military might were seeping through the historical cracks, and beginning to grip world events.

The Great War of 1914 unleashed all the old, ugly patterns of warring powers and added new ones, perhaps uglier still, of nationalism and ideology. In the run-up to the war, the balance-of-power strategy, designed to gain advantages for some and deny them to others, spun out of control. Millions of Europeans were killed, and dynastic emperors were forced from their thrones (and even their countries), thus opening the floodgates to democracies and modern dictatorships, as well as to worldwide economic and political instability. The carnage and costs of World War I were staggering.

With the great powers of Europe exhausted, it seemed that the players of the power games would be forced to change. The Austro-Hungarian, Romanov, Ottoman, and German empires lay vanquished. France, Italy, Japan, and the United Kingdom were the major colonial beneficiaries. But Europe's enthusiasm for war had dimmed with the vast numbers of dead in France and the United Kingdom. Czarist Russia became Lenin's Soviet Union and started on the path to major power status all over again, under communist rule. Germany surrendered, but was never actually conquered and was steeped in a passion for revenge. None of the European continent's major players were in a position to fight big wars for at least a dozen years. This gave some of their colonies room to breathe, to nurture hatred of their occupiers, to plan resistance, and to dream of creating their own new political entities.

Woodrow Wilson, the president of the newest and one of the strongest world powers, was determined to set firm limits on international power, especially that to make war. He assumed center stage to lead the exhausted nations of Europe into the next world—a world where aggressors would have to think thrice about the consequences of aggression, and where new kinds of states would ultimately reject war. He would do this in two ways: first, by substituting collective security against aggressors for the undependable balance of power; and second, by reducing aggressive empires into what he hoped would become peace-loving, democratic entities of uniform nationalities. Machiavelli might have labeled him a "prophet armed"—not a compliment from a writer who worried that such prophets might run amok. In Machiavelli's day, just such a prophet-priest, Savonarola, took over Florence, eventually consuming the city in his religious madness.

Wilson's dream that World War I would be the war to end all wars did not come to pass. Europe's major powers were not ready to practice collective security. Nor was the United States ready to shoulder new peace-enforcing responsibilities. Wilson's belief in the restraining power of national self-determination and democracy failed to rein in the long-held ambitions of the major powers.

Hitler's Germany and Hirohito's Japan scrounged all their countries' resources to transform themselves into unstoppable military dynamos. With military power and will, they could do almost anything they wanted. But for the luck and will of their adversaries and their own blunders, these dictators might have succeeded, at least for longer than they did. In many ways, the first half of the twentieth century represented both the high point of the old order of power—minimal checks on the power of the strongest—and the beginning of its demise.

FIRST CAME THE AWARENESS among the major powers of new dangers and of the need for new caution. The blockade of Berlin in 1948 set a pattern: mutual Soviet-American testing and threatening, but avoiding, at all costs, going to war. The Soviets blocked American and Allied land traffic from the divided city. At a decided military disadvantage, the West responded with an airlift and kept its part of the divided city alive. Soon enough, Moscow reopened the roads to West Germany.

Shifting the testing ground to Asia, which seemed less dangerous than Europe, Soviet leaders gave the green light for North Korea to attack South Korea, and later for Chinese intervention as well. Korea could have escalated into a Soviet-American war, but Moscow and Washington purposely avoided it.

The Cuban missile crisis of 1962 was even more dangerous. Moscow lost its bearings and secretly placed missiles in Cuba; Washington announced it would not tolerate them. The United States imposed a naval blockade on Cuba to prevent additional missiles from entering the country. Moscow did not challenge the warships. The one and only direct Soviet-American crisis concluded with pledges of mutual restraint, agreement on a nuclear-free Cuba, and a secret arrangement to withdraw U.S. missiles from Turkey.

From then on, the two nuclear-armed titans treated each other with the greatest care, never again even approaching direct blows. Their rhetoric was sometimes extreme, as befitted deeply opposed ideological foes, but was used mainly to justify a continuing arms buildup on both sides. Known as the arms race, it produced constant tensions, but little more. Less confrontational and less dangerous proxy wars in the Third World were as far as the United States and the Soviet Union would go toward escalating conflicts, which was not very far.

Never before had two great powers with such profound conflicts of interests and values refrained from direct combat. Imagine Athens and Sparta without the Peloponnesian War, Rome and Carthage without the Punic Wars, or France and the United Kingdom without each other as an enemy. The final acts of the Cold War were played out not on battlefields, but in the doleful decisions of Communist Party leaders in the Kremlin's Politburo meeting rooms. The nearly fifty-year struggle between history's two titans ended not with a bang but a whimper, and without a clash of their unmatched armed forces. The Soviet Union had self-destructed. America stood alone in the ring.

THE SUPERPOWERS HAD UNDERSTOOD the first new law of power and sidestepped direct confrontations, but both—as well as other major players such as the United Kingdom and France—still failed to appreciate the decline in their own power over territories overseas. As empires fell and colonies became new states, as Soviet-American competition moved through the Cold War, and as the two superpowers tried to control the nations in their spheres of influence, they all eventually ran into brick walls. They all shared a difficulty coming to grips with the fact that the new emerging nations were born out of resistance to their former colonial masters and would, above all, know how to resist in the future.

Soviet leaders discovered this new wall of nationalism early in the Cold War. In 1948, Marshal Tito (Josip Broz), the communist leader of Yugoslavia, defied the general secretary of the Communist Party of the Soviet Union, Joseph Stalin, the leader of the most potent military by far in all of Europe—and got away with it. Stalin wanted Tito to toe Moscow's line. But Tito had popular backing and rebuffed Stalin's orders. Stalin threatened and made life harder for Tito, but essentially let the matter fade away.

The United Kingdom seemed to get the message about the new limits on power when, in 1947, the British Parliament accepted independence for India (and Pakistan), the jewel in the crown of its empire. Kicking, stalling, and sometimes fighting, the British granted independence to their other former colonies, one by one, over the following decades.

Whereas the United Kingdom was realistic about its colonies' desire for independence, France was stunned when, on May 7, 1954, General Vo Nguyen Giap and his Vietminh forces received the surrender of French forces besieged at the town of Dien Bien Phu in the French colony of Indochina. The French and Vietminh forces had fought each other in that valley for fifty-five days. At the end, France was drained of its will to fight to retain control of Indochina.

For arguably the first time in modern history, a rebel army without a state had defeated the army of a major power in pitched battle. Ho Chi Minh and his army had rallied the new forces of anticolonialism and nationalism and had won the backing of the Vietnamese people. With foresight, French leaders could have read the tea leaves and begun dismantling their empire, but they were not yet ready to do so. They would fight and lose for years in Africa.

Then, in 1956, came the Suez Canal crisis. The United Kingdom joined with France and Israel to invade Egypt to gain control of the Suez Canal and to teach a lesson to the upstart leader of Egyptian nationalism, Gamal Abdel Nasser. But Washington and Moscow would have none of it. President Dwight D. Eisenhower threatened his own allies with harsh economic penalties. London and Paris withdrew their troops and suffered a deep humiliation. Suez was a watershed. To future Nassers, it delivered the message that the world would back them against their colonial masters. To leaders of colonial empires, it warned that their allies would not support large-scale colonial wars.

Not to be forgotten is that when Nasser overthrew King Farouk in 1952, it was the first time since the sixth century BC that an ethnic Egyptian had ruled Egypt. This was to be a prophetic event, as other national leaders gradually assumed power over their own people, replacing local maharajahs, tribal chiefs, and foreign viceroys who were relics of the Ottoman, British, and other empires. And with independence and self-rule came the will to resist both foreign military occupation and pressure.

In this postcolonial world, the United States started off with everything on its side. It had a generally good record of opposing colonialism. It was prepared to help new states maintain their independence through the new United Nations, economic aid programs, and financial institutions such as the World Bank. To most of the newly independent states, the United States, itself a former colony, represented freedom and opportunity. But Cold War competition and Americans' fears of communism soon overrode these early impulses. Thus began Washington's travails in the Third World.

Washington saw the Soviet hand behind almost every Third World upheaval. And so it overthrew freely elected governments said to be pro-communist in places such as Guatemala and Iran, unwittingly planting the seeds of future troubles. Washington also supported many dictatorships that seemed the only alternative to communism. In general, American policy was to step into the vacuums created by departing colonial powers and fight the potential extension of Soviet power there. But alongside these invariably sad stories, the United States did have genuine success in helping South Korea and Taiwan evolve into thriving democracies.

The most dramatic setback for the United States, of course, occurred in the very place where France had been brought to its knees—Vietnam. Here was America's first deadly taste of the limits on its power. And North Vietnam won. Nixon proclaimed that the United States would become "a pitiful, helpless giant." To this day, many argue about whether Washington could have won the war by using greater force and staying the course. But even had the United States prevailed on the battlefield for some time longer, it seems certain that Americans would have tired of fighting this new "television war" (the first American war shown on TV at dinnertime every night), and that Hanoi would have licked its wounds, resumed the fight, and eventually won. The war opened a debate on the limits of U.S. power that still resonates. But the fact remains: a tiny, backward fragment of a nation called North Vietnam prevented the mightiest military power in history from having its way.

A second blow soon struck the United States, this time from the other end of Asia, in Iran. In late 1979, Islamic revolutionaries took fifty-two Americans hostage and held them in the U.S. embassy compound. It was an outrage, condemned almost universally. President Jimmy Carter's mismanagement of the crisis, however, deepened America's shame and image of impotence. Most of the Americans were held hostage for 444 days, and became a daily reminder of how little power the greatest country appeared to have over a lesser one.

In 1979, the Soviet Union confronted its own Vietnam in Afghanistan. Soviet troops poured into that country to prop up a communist government, fearing it would otherwise succumb to Western influence. The vaunted Soviet army lost to Islamic and tribal guerrillas, who relied on Western arms and their own hatred of foreigners. Soviet forces, considered among the world's best, withdrew in 1989. The Soviet empire in Eastern Europe fell to street demonstrations later that same year, and in 1991, the Soviet Union itself vanished.

By the end of the Cold War, it might have appeared that the world's two major military powers had learned some lessons on the limits of their armed forces—though time would prove otherwise. But it remains beyond doubt that the great powers had learned to fear war with each other. They absorbed the lesson of this new danger so well that they refrained entirely from attacking and destroying one another—at least, directly—for more than six decades, from 1945 to the present and still with no serious threats on the horizon. This may be the longest stretch of peace among major powers in modern history.

BUT ONE OLD PATTERN that had not outlived its usefulness was the balance of power. Throughout the Cold War and beyond, many nations sought the United States as a balancer against regional threats. And sometimes, they looked to one another and to other major powers to keep the United States at bay.

Both Eastern and Western Europeans remained close to the United States as a counterweight to Russia. Moscow was still the big neighborhood bear. Similarly, Asian nations wanted American presence and power to check what they saw early on as a rising China. The fact was and remains that they view Beijing as a far greater threat than Washington. Similarly, Middle Eastern countries have clung to Washington, to protect them first against Saddam Hussein's Iraq and later against Iran.

The fact that so many countries continued to play the balance-

of-power game testified to the continuing centrality of power—
whatever leaders of these nations said to the contrary. No changes
in the patterns and laws of power have altered the advantages to
countries of playing off friends against foes and on occasion, even
friends against friends.

Today is perhaps the first time in history that relationships and
issues among the major powers—the struggle for an international
balance, advantage, or hegemony—are not the be-all and end-all
of the international system and its agenda. The focus today is not
on the major powers' competition and conflict, but on nuclear nonproliferation, terrorism, climate change, energy, and a wide variety
of international economic issues. How relations develop among the
United States, China, Russia, India, and the major Europeans is, of
course, critical, but the current level of competition and conflict is
relatively low by historical standards. This, in turn, places a particular burden on the world's major powers to keep conflicts among
themselves tamped down.

As THE WORLD ENTERED the 1990s, the United States loomed as
the sole superpower and probably the strongest power in history,
both absolutely and relatively. In military, economic, and technological power, it dwarfed both its contemporaries and all empires
of the past. No other nation had ever equaled America's military
might, its destructive power, or its enormous and technologically
advanced economy and its worldwide reach. Never had the gap been
greater between the power of the leading nation and all others. No
nation occupied even a close second place. Rome had Carthage and
the Germanic tribes. The Ottomans had the Europeans. Napoleonic France had the British and eventually many other Europeans. While the United Kingdom dominated the world's oceans, its
land power was exceeded by Germany's. During the Cold War, the
Soviet Union and America had each other. Perhaps only Rome for a
while and China for somewhat longer ever ranked so far above their
contemporaries, but neither ranged the globe.

The First Iraq War seemed to confirm America's dominance. President George H. W. Bush amassed allies and ran Saddam's forces out of Kuwait quickly and decisively. It seemed that in the post–Cold War world, Washington could do as it wished. The good old days of the strong commanding the weak were back, not as tragedy or comedy, but as a heroic era, or so it seemed. There was no nation or bloc to counterbalance the United States.

Trouble brewed in the Balkans, but the first President Bush did not regard this as a test of American power and would have no part of it. With the breakup of Yugoslavia, the Serbs' killings of Bosnian Muslims and Croats mounted. Europe and the United Nations did nothing of consequence in response, and it began to look as if in the American era, genocide would be tolerated.

The sense that America was powerless or uncaring in the face of such humanitarian disasters ballooned further as Somalia descended into civil war and mass starvation. At first, Bush did nothing; then, in 1992, he joined a UN relief effort with a small military force. President Bill Clinton expanded the American presence, got involved in the civil war, lost a handful of troops in an ambush, and then ignominiously pulled out the American contingent in 1994. Growing whispers of America's lack of will were somewhat blunted by the strong action Clinton eventually took in Bosnia and with the Dayton Accords to stop the carnage there.

The terrorist attacks on September 11, 2001, were the next milestone. America suddenly sensed itself as vulnerable to three new threats—terrorism, failed states, and weapons of mass destruction, particularly nuclear weapons. And two states, Afghanistan and Iraq, quickly came to exemplify these threats. President George W. Bush would go after both in force, but the nation, if not the president himself, would soon be reminded of the limitations on American power in such ventures.

At first, Bush did the right thing in responding to the terrorist attacks by ousting Afghanistan's Taliban government, the shelter of the al-Qaeda terrorists who had attacked America. Within weeks, the Taliban had fled from the capital, Kabul, to the mountains with al-Qaeda. In his January 2002 State of the Union Address, he took on the "axis of evil"—Iraq, Iran, and North Korea. These were now his main targets. In the meantime, he failed to destroy the Taliban and al-Qaeda in Afghanistan, and by 2006 they were making a comeback. Washington once again found itself mired in an inconclusive war in yet another backward country.

On March 20, 2003, Bush attacked Iraq because, he insisted, Saddam had weapons of mass destruction and was deeply involved in terrorism. Within weeks, U.S. forces had smashed and overthrown Saddam, and had taken Baghdad. Then, as always, mistakes were made: Bush failed to provide sufficient forces to maintain a minimum of security, Iraqi forces were dismantled, and the U.S. occupation authority bungled the political situation.

As in Afghanistan, a formidable insurgency arose, this time composed of Saddam's defeated loyalists, foreign fighters from al-Qaeda who rushed into Iraq to assist them, and Shiite militias. To add to Washington's woes, Bush found no weapons of mass destruction, thus undermining the principal rationale for the war and laying the basis for a shift in rationale to the far more demanding goal of transforming Iraq into a free-market democracy. At an even greater cost in blood and treasure than in Afghanistan, the fighting dragged on for several years with no clear end in sight. Inevitably, early popular support for the war disintegrated into debates, doubts, and demands to withdraw, and then settled down in 2008 as fighting and casualties declined. Once again, the war's supporters had to confront the central lesson of the last fifty years—that military power alone, no matter how decisive against conventional armies and entrenched forces, has sharp limits. It required political compromises and arrangements with former Sunni and Shiite foes to really begin to turn the situation in Iraq around, or seemingly around, by 2009.

As for Iran and North Korea, the other states in the axis of evil, both went on to increase their nuclear capabilities, despite threats from the Bush administration. North Korea exploded a nuclear device, though at one point it agreed to limit its nuclear programs in exchange for economic aid and security guarantees from the United States, in a deal that collapsed in late 2008. Into the Obama administration, Iran continued with its uranium enrichment program, and North Korea continued to explode nuclear devices.

As for Afghanistan, America's initially decisive military charge was soon blunted, and U.S. and NATO forces there became mired in the anti-Taliban war. America's long and costly military efforts in Afghanistan and Iraq were coming to have the paradoxical effect of *undercutting* the prospect of America's using military power anywhere else—against Tehran or Pyongyang, for example. To state the paradox starkly: the American military had easily defeated the armies of the Taliban and Saddam, only to get bogged down in insurgencies, then to see support for military force erode at home, resulting in increasing doubts abroad about the credibility of America's military power. At this point, the Iraqi insurgents, the Taliban, Iran, and North Korea have all defied U.S. power and either gotten away with it or made America pay a very high price.

MILITARY POWER STILL MATTERS, and sometimes it matters a great deal, as it did in the First Gulf War and the Balkans, as well as in moving the Libyans toward nuclear arms negotiations in 2003. But it isn't what it used to be. Historically, military force often had a clarifying effect on relations between states; it definitively or mostly settled matters one way or another. Leaders knew what to expect. They knew that the country with superior force would almost always prevail. But as the world moved toward the twenty-first century, that certainty began to dim, sometimes sharply. Major powers could no longer count on the success of force as they did in the past. The utility of force had been muddied, and that severely complicated calculations of the uses of power.

Further, as the efficacy of military power waned throughout the latter half of the twentieth century, economic power—or the ability to alter policies in return for trade, aid, expertise, and investments—began to wax. This change, too, would rearrange patterns of power and place new constraints on the major powers. It would become at once a power tool of the economically mighty and a new constraint on their use of power.

As the twentieth century reached its end, the strong found themselves running into an ever-expanding set of complications and restraints on their power: the irresistible force of nationalism, the multiplying number of new states, new international norms and institutions dedicated to protecting the emerging states, the revolution in worldwide communications, internal political and economic demands within the major powers, new kinds of threats, and new, and less easily settled, conflicts rooted in ancient disputes within the new states.

Nationalism had made the weak better able to resist the strong. Big powers easily pushed around local tribal and traditional leaders, most of whom did not command much loyalty from their subjects or many resources. But in the mid-twentieth century, peoples around the world increasingly sought and won their independence and the right to establish their own states. Formerly submissive peoples were increasingly prepared to sacrifice countless lives and do whatever necessary, and for however long it took, to win and maintain their independence.

The number of new states increased dramatically. Initially, there were thirty-two members of the League of Nations. The original membership of the United Nations tallied fifty-one. The number rose to 144 by 1975, and then in 2006 topped off at 192. To be sure, very few of the new states encompassed just one coherent ethnic, religious, or tribal group; almost all were mixed. Their mixed composition virtually guaranteed either internal conflicts or repression, civil wars, and sometimes breakaway—and ultimately even newer breakaway—nation-states.

The new states also discovered additional protection against the strong in norms, international law, and international institutions. They cloaked themselves in the legal mystique of sovereignty, the centuries-old doctrine of international law that established rights for such entities to exist and to be supreme within their own borders.

Having enjoyed the blessings of sovereignty for centuries, major countries were in no position to deny these rights to the newcomers.

The United Nations also tended to enshrine sovereignty. The United States, notoriously protective of its own sovereignty, generally backed UN norms of sovereignty. There were occasional exceptions, as when Washington and others were particularly appalled by state violations of human rights and helped pass appropriate UN resolutions calling for international corrective action. And UN Secretary-General Kofi Annan promoted a new doctrine of humanitarian intervention when regimes abused their people or committed genocide. But these humanitarian views did not prevail.

The Security Council, the key institution within the UN on security matters, became an additional restraint on the use of military power, especially for the United States. Weaker nations, joined by many Security Council members such as Russia, China, and France, viewed the Security Council as a means to check American power and sought to strengthen the Council as the world's only source of legitimacy for the use of force. The Council only rarely and reluctantly provided this blessing. Over the years, the Council approved only a handful of military interventions, although it often assented to peacekeeping operations. The overwhelming majority of UN members regularly and strenuously resisted the use of force against any member state, almost regardless of its transgressions, internal or external, for fear that the ax might one day fall upon them for similar wrongs.

The growth of international news media with instant global reach also impeded the exercise of force by major powers. Far more than ever before, information was instantaneously available, and this phenomenon gave rise to the almost equally immediate formation of world opinion, usually in opposition to strong-arm tactics by major powers. In earlier times, colonial troops could kill thousands in distant lands with little public awareness. Now, the striking of a single home by a bomb and the starvation of children caused by economic sanctions fill television and computer screens around the globe within hours, if not minutes.

Weaker states have also found relatively inexpensive ways to oppose superior military force. These force equalizers, as they are called, include guerrilla warfare and terrorism, as well as the threatened use of rudimentary weapons of mass destruction. Many such states already possess the technological prowess to make chemical, biological, and even nuclear weapons. These programs are usually difficult to detect, and therefore difficult to eliminate. Of perhaps greater significance, major powers have become acutely aware of their own vulnerability to attacks by weaker states and terrorists armed with weapons of mass destruction.

Major powers also find themselves restrained today as never before by their own internal politics and economic needs. As recently as the 1930s, even the most advanced industrialized democracies paid only modest attention to the health, education, and welfare of their own people. Now, these needs compete for resources against defense budgets. In Europe, guns lost out after the Cold War to butter, but not in the United States.

Great powers are also challenged by many twenty-first-century threats—threats that now stem far more from within nations than from between them. Before, international power was applied mostly to actions states took beyond their borders against other states. Now, the threats often arise from states harboring terrorists, developing WMDs, or simply falling apart and becoming failed states. Countering such threats often entails the costly and burdensome task of occupying a country rather than the simpler one of defeating its army on the battlefield.

America's power in the world has been further tempered by the achievement of its own international goals: the development of democracies, free markets, free trade, international law, and national self-determination. America championed these ideas, and now, ironically, they have returned to restrain America itself.

THIS STORY OF THE great American Gulliver bound by Lilliputians, of increasing limits on great powers since World War II, has attracted many authors. In the 1970s, Professor Stanley Hoffmann of Harvard wove the case brilliantly around the explosion of nation-states, nationalism, sovereignty, international law, and norms. In the late 1970s, Professors Joseph Nye and Robert Keohane, also of Harvard, demonstrated how new economic and political interdependence added to these restraints.

Why revisit this oft-recounted history? Because as many times as it's been told, it hardly dents the conduct of American leaders. Hard-liners keep ignoring, forgetting, or simply disregarding the existence of the limits on power, and they keep saying that, with the proper national will and vision, the United States can do anything. And Americans keep falling for this rhetoric. Sometimes gentler souls appear to commit the opposite error. They say that the limits on power are so great as to make America virtually powerless, and argue that Washington can protect itself only through understanding, love, and dialogue. Both are wrong.

Sometimes, the hard-liners accept the limits of power in theory, only to deny their existence when it really counts. This happened repeatedly in recent decisions on Iraq. Conservatives were skeptical about America's capacity to engage in nation-building, yet they tried to do just that in Iraq and Afghanistan. Sometimes, they will grant the general idea of limits, but then deny its application to the case at hand, such as making idle threats against Iran. And then there are those who claim they understand the limits on the use of force before they get into elected office, only to ignore these limits once in office and under fierce political pressure. Clinton did this when he expanded a humanitarian relief operation in Somalia and entered into that country's brutal civil war. Sometimes, leaders privately get the picture, but deny it publicly because they think that acknowledging limits on American power is bad politics. In 1993, Clinton nearly fired Undersecretary of State Peter Tarnoff, who had spoken of the limits of American power in a chat with reporters. A number of his superiors, it seems, worried that the very suggestion of limits exposed the administration to charges of being too liberal and therefore unwilling to exercise American power. In Washington politics, it is easy to confuse limits with impotence. Accepting limits is equated with weakness. It is considered virtually un-American to argue that the United States might be incapable of doing whatever it wants.

Sometimes, even those enlightened about restraints lose their footing in the pursuit of great causes, especially their own. Liberals, and now conservatives as well, insist that America must employ its power to democratize the world. Many liberals joined Bush in his initial push to bring democracy to Iraq and the Arab world. After all, the United States had won the Cold War and was now riding the crest of history. Now, many announced, was the moment for great deeds.

Americans continue to err in two basic ways in thinking about their present-day power: First, some deny limits altogether and cling to the grand fantasy of American omnipotence. Second, others embrace limits and assert American impotence. Many conservatives commit the first error, and many liberals commit the second. But the United States is neither omnipotent nor impotent. The internal and international constraints on power cannot be dismissed, nor are they insurmountable.

From the birth of the nation-state until after World War II, international power was much simpler. It was about the threat or use of force, and military superiority worked far more often than not. Military power was the lubricant of international affairs, solving some problems while creating others. Now, in a world filled with constraints, exercising power is a more complicated business—as exemplified by the continuing survival of communist Cuba ninety miles from the shores of the greatest power in history.

American leaders have to stop either blithely ignoring or driving directly into international minefields. These snares and traps won't vanish. They are now stubborn, permanent features of the international landscape—and while they restrain power, they do not shackle it.

American leaders can learn to work with and around these limits and thereby enhance American power in the twenty-first century.

They have to start by retrieving the real meaning of power, a meaning lost in the intellectual and political morass of the struggles between conservatives and liberals. They must also come to terms with the new distribution of international power, which will reveal both what's achievable and how to achieve it.

CHAPTER 2

WHAT POWER IS, AND WHAT POWER ISN'T

In Shakespeare's Henry IV, Part I, *the warrior and braggart Glendower proclaims to the heroic Hotspur, "I can call spirits from the vasty deep." To which Hotspur retorts: "Why, so can I, or so can any man; [b]ut will they come when you do call for them?"*

Like Glendower, U.S. leaders have done a lot of summoning without anyone's coming, in good part because they don't understand the core meaning of power.

Political leaders and policy experts know that power is the platinum coin of the international realm, and that little or nothing can be accomplished without it. They know that whoever wins the battle to define power in Washington controls U.S. foreign policy. So, they fight on, and often bitterly.

Entries in these lexicographical tussles include the power of ideas, the power of leadership, the power of personality, the power of persuasion, the power of values and convictions, the power of example, the power of the purse, and of course, the power of brute military force. In recent years, participants in these tussles have also spoken of hard power, soft power, smart power, and—when referring to their opponents—dumb power. These many choices conjure up all the intellectual joys of a menu in a Chinese restaurant.

There are the liberal offerings that tend to define power as based on an understanding of others, leadership, communications, principle, values, persuasion, and reason. Liberals usually regard pushing others around as excessive and self-defeating, or as a bad last resort. It's almost as if they believe that Washington possesses international power to the extent that it can convince others that America's wishes and values match or should match their own, and that rightly understood, almost all differences can be resolved by reason. Liberals will be shocked by the starkness of this description, but in my experience both in and out of government, this is basically how they view power in their heart of hearts.

American conservatives tend to emphasize national self-interest over universal principles (although Reaganites and neoconservatives emphasize principles as well), and pressure over persuasion. They believe that the American way is the right way, that America's adversaries are either wrong or evil, and that more often than not, being reasonable and negotiating plays into villainous, untrustworthy hands. To deal with an unruly and threatening world, they define power mainly as military might and the fear its use can engender in our enemies. Unlike liberals, in my experience, conservatives are not embarrassed by this hard-nosed image of themselves; indeed, they embrace it.

Moderates typically offer a little from column A, a touch from column B, and a pinch of C—without an overall ideology and without much effect on the political debate. Moderates generally don't fare well in the political trenches; their distinctions and subtleties fade in the fray. They generally shy away from overall strategy, although, to their great credit, they emphasize facts and sensible pragmatism. I've watched them lose countless political debates—initially and for a while. They are not good at packaging their ideas for popular consumption; nor are they good proselytizers. They're too complex and don't know how to make complexity more simple. However, years down the line they often win by default—after years of costly policy failures caused by the more dogmatic liberals and conservatives.

Arguments about power have been the stuff of presidential campaigns as well as fierce battles waged on op-ed pages around the land, and the latter are not merely examples of rhetorical jousting: Jimmy Carter accused Henry Kissinger (and through him, Richard Nixon) of sacrificing true American values on the altar of balance-of-power cynicism (although when I interviewed Carter off the record as a reporter for *The New York Times* at the end of the 1976 Democratic primaries, and asked him who he thought was the nation's most able foreign policy thinker, he replied, "Kissinger"). To Carter, values were power. Ronald Reagan charged Jimmy Carter with exalting vague ideas of human rights over national interests and of being naive and weak when it came to Soviet power grabs. To Reagan, clear military superiority was the heart of power, although he later added values to his quiver. Bill Clinton pounded George H. W. Bush and Bush's secretary of state, James A. Baker III, for heartless and un-American realism in the face of threats to humanity in the Balkans and in Africa, yet he himself did nothing in either arena for years. To Clinton, international power was mainly economic power. And George W. Bush attacked the Clinton team for draining America's military strength in its vain attempts at nation-building and for its weak responses to dictators like Saddam Hussein, yet Bush undertook the two biggest nation-building enterprises in American history. To the second Bush, the will to use force and to stay the course was power. However intellectually flawed their concepts of power may have been, these concepts helped each man to capture the White House.

The cycle of winning and losing definitions is familiar: Force proves too costly or ineffective to one president, and a rival unseats him with pleas for dialogue and values, and vice versa. At the end of the day, both conservatives and liberals actually do harm to U.S. foreign policy by misconstruing what they cherish so dearly—the nature of power itself.

Power is getting people or groups to do something they don't want to do. It is about manipulating one's own resources and position to pressure and coerce psychologically and politically.

It's easy to confuse power in personal relations or within nations and power between nations. With people and inside countries, power can often be exercised by personality, leadership, values, persuasion, and the like. There are generally far more shared values and interests in these venues than in international affairs, and therefore give-and-take can be relatively contained and controlled and thus be quite effective. Between states where there is little common framework, power necessarily takes on a harder edge and is more difficult and complicated to apply. It's easier to say no. In international affairs, persuasion and leadership rarely work, and force can be met with counterforce, especially today. In twenty-first-century international relations, power entails far more pressure and coercion than it does in domestic politics or personal relations.

LONG BEFORE THE WORLD evolved to its present state and even before the dawn of think tanks, the ancients perceived the distinct nature of international affairs. The Chinese and Roman empires lasted far longer than all those that followed, in both cases because their leaders fully grasped the richness of power and the wide array of instruments and techniques for exercising it. They were particularly adept at using their superiority in resources and generating successes to intimidate and co-opt adversaries. This, in turn, allowed them to avoid the constant application of costlier forms of power, such as military force. China and Rome accumulated immense resources and power, and never lost an opportunity to remind everyone else of their superiority. Both, nonetheless, refrained as much as possible from pounding their conquered subjects militarily, and this restraint led both empires to rule relatively effectively and cheaply. The United Kingdom, more than other later empires, copied this strategy well.

No one thought harder or with more insight and sophistication about power and governing than the ancient Chinese and the Romans. They saw power from many angles: as the potential to do damage or confer benefits, as momentum toward future accomplishments, as the grandeur of legitimacy and authority, as reputation, as cultural superiority, and as brute force. They calibrated their use of power to fit each situation, understanding that they had to treat some more subtly than others. But most important, they rarely lost sight of power's psychological and coercive essence, and of the skills required to wield it. Alas, Chinese thinking had little impact on Western thinking until the nineteenth century.

The Romans figured out most of this on their own, always with assists from ancient Greece. Like the Chinese, the Romans shunned a single word for power, for controlling the spectrum of political behavior. The Romans, also like the Chinese, knew that military force underpinned all power, and probably relied on it more than the Chinese. Rome, as well, went out of its way to exercise power in a myriad of forms so that its empire would be spared the overbearing costs of war as much as possible. One figure shows this determination starkly: Rome's own military manpower peaked at a mere 375,000 legionnaires to provide security for an empire that stretched from Britain to Persia. And to make sure that force would always be a last resort, Rome usually gave wide political latitude to most of its conquered dominions; it ruled essentially through the authority and prestige of Roman officials, through the acceptance of Roman law, and by inducing compliance with its laws through clever offers of citizenship and honors to its subjects.

Western heirs of the Roman tradition couldn't quite settle on precisely what power meant. In English, they used it mostly as a noun, though its etymological roots clung to the Latin verb *potere*, "to be able to." Romance languages adopted variations on the Latin, but kept its meaning. In Spanish and Portuguese, it became *poder*; in Italian, *potere*; and in French, *pouvoir*. According to the *Oxford English Dictionary*, the word first made its way into Middle English in 1297 as *poer*. In the fourteenth century, the *w* was added to spell *power*. But throughout, it was understood to mean *ability* as a noun and *to be able* as a verb. In the fifteenth century, its meaning took flight in the form of images like God's power, nature's power, and spiritual power. Perhaps the authors of the time were seeking to convey a phenomenon beyond human abilities, thus giving the word itself mysterious qualities.

In general, Western writings on power were vague and unspecific, as they are today, and were less meditations on its meaning and more tributes to its prowess or religious condemnations of its corrupting attributes. One man, however, was to describe it both sharply and enduringly.

It was Niccolò Machiavelli who, in *The Prince* in 1513, put meat on the vague bones of *potere*, giving it specificity and tangibility—without much mention of the word *power* itself. To him, the art of being a prince, or of wielding power, meant being able to protect the state, to make alliances as necessary, and to conquer and rule other states. To do this, Machiavelli advised the prince that all he had to do was to learn the art of war. For a prince to do what he must, his state needed both good weaponry and well-trained soldiers. A full treasury was also helpful. And the less dependent his state was on another in these martial endeavors, the more power he would have. Great power, in effect, amounted to self-sufficiency in warfare. "A prince," Machiavelli wrote, "should have no other object, nor any other thought, nor take anything else as his art but that of war and its orders and discipline, for that is the only art which is of concern to one who commands."

Apparently, Machiavelli felt no compulsion to actually define power, because its essence seemed obvious to him and his contemporaries—namely, to maintain strength and unity at home, to successfully defend one's own turf, and to be able to threaten and successfully wage war. These dicta on the relationship between ruling and war brilliantly foreshadowed the actions of kings and queens, princes and presidents for the next 450 years.

Over the centuries, *The Prince* morphed into *the* power cookbook. (Critics ignored Machiavelli's more extended thinking in his *Discourses on Livy*, which placed ethics and good behavior at the center of political life for both ruler and ruled.) Through most of European and American history, it was popular to condemn Machiavelli's precepts of power—only to then follow them. There wasn't much competition, save for Thomas Hobbes in the 1700s. Most other philosophers must have seen the subject as somewhat immoral and even un-Christian—or they feared that the Florentine master had already said all that was necessary on the subject. Machiavelli's rendition of power towered over others in the West: Power was military superiority, force, and the convincing threat of force.

While doing little to illuminate the literature of power, the British nonetheless conquered and preserved an empire that reached from the English Channel to Africa, India, Australia, New Zealand, and even a foothold in China. Of course, the British army and navy were ever-present threats to native rulers, and locals could never quite rid themselves of their fear of the "great beef-fed men, red-faced and red-coated" that George Orwell described in *Burmese Days*, as British troops marched through the streets of Mandalay. But equally vital to the long and successful rule of the British was their ability to co-opt local rulers through offers of wealth, through trade, by means of Christianizing and anglicizing them, and by giving them British common law. Like the Romans' lure of being a Roman citizen, the British would seduce conquered local rulers with the intoxication of being deemed British gentlemen.

Other major modern empires—the Soviet, German, and Japanese—were relentlessly ambitious, relying almost entirely on military force and fear, and didn't last long. The Soviets held on longer than the others partly because communist ideology held a certain appeal in some countries for a time.

PROBABLY THE BEST DEFINITION of the basic workings of power, although not the most eloquent, was proffered by the great political scientist Robert Dahl of Yale, who wrote in a classic 1957 article, "The Concept of Power": "*A* has power over *B* to the extent that he can get *B* to do something that *B* would not otherwise do." This occurs, in Dahl's explanation, when *A* actively pressures *B* or because *B* anticipates and therefore fears the pressure. The key is pressuring *B* to act against his will or desire, i.e., to change his behavior. By pressure, Dahl certainly did not mean persuasion, which is essentially an intellectual or emotional process. Nor did he suggest that he meant force, which is a physical act.

Power is mental arm wrestling. It derives from establishing psychological and political leverage or advantage by employing resources (wealth, military capability, commodities, etc.), position (such as a geographic regional balancer, or a political protector), as well as maintaining resolve and unity at home. These are embodied in a process whereby *A* convinces *B* that *A* can and will help or harm him, give him pleasure or pain, relieve his difficulties or increase them—whatever the costs to *A* himself. Power thus varies with each and every relationship and changes with each and every situation. It has to be developed and shaped in almost each and every situation, and will vary over time and place. And critically, the wielder of power must take great care to be credible, to be taken seriously, both at home and abroad.

Power-wielders gather their resources—economic and military assets, world and domestic opinion, national unity and resolve, and their international position—and package them into words and deeds. By these words and deeds, they create impressions about the likely consequences that others will experience, as well as offer foretastes of their potential losses and gains. Rightly designed and exercised, power reaches into another's society and government in order to strengthen allies and weaken adversaries. The successful exercise of power requires great skill and art. Luck also helps.

Even the greatest artists must first accumulate mundane materials such as good paint and brushes, or clay, and so it is with power. Nations cannot play the power game without the requisite resources, and the greater the power the more resources it requires. There will be no more Macedonias—poor states with a mythic Alexander the Great—that can conquer the world. Since the rise of the modern nation-state in the sixteenth century, power has rested on a strong economic base, on a firm command of military technology and weaponry, on a united popular will, and, often, on strong leadership. From these resources, great power-wielders create their power.

But a power base is much more than simply adding up resources: it depends on the kind and nature of these resources—namely, a nation's relative self-sufficiency and resilience once a power struggle begins and the squeezee squeezes back. Resources measure only static assets, when what's critically important is how those assets hold up once the arm wrestling begins. Thus, when Arab states cut back on global oil supplies in 1973, sending prices skyrocketing, in order to force Washington to press Israel for concessions, the United States had both sufficient oil reserves and a leadership capable of bracing Americans to pay higher gas prices.

Power is also relational. It is not fixed between one nation and all others. It depends on the exact relationship of the countries involved. Generally, the relationship hinges on two factors: the static and dynamic resource gap, and the kind of relationships forged by the nations' leaders. Obviously, American leaders have much more going for them vis-à-vis Nicaragua and Thailand than vis-à-vis India or Russia. Or take the relationships of British and French leaders with George W. Bush. The United Kingdom and France are roughly equal in resources, but Prime Minister Tony Blair's support for Bush, compared with President Jacques Chirac's critiques, put Blair in a much stronger position to influence Bush.

Power is situational as well, meaning that it depends on the exact circumstances once the pulling and tugging begin. Saudi Arabia has sway in Washington because of America's oil dependency and because of Saudi investments in America. But when threats to Riyadh arise, as when Saddam invaded Kuwait in 1991, the Saudis' need for American protection became the paramount consideration. The Saudis even paid a large chunk of the bill for the war. Another recent example is the relations between Uzbekistan and other "stans" (the new nations of Central Asia, formerly Soviet republics) and the United States at the start of the Afghan War in 2001. The stans needed Washington as a balance against Moscow and Beijing, and Washington wanted their oil and gas. But when George W. Bush decided to retaliate against the Taliban and al-Qaeda as perpetrators of 9/11, and by invading Afghanistan, the balance tipped toward the stans. Bush needed air and logistical bases in the stans to fight the war, and in return, he muffled his criticisms of their dictatorial leaders.

From this, it's also apparent that power turns on one's position in regional or international affairs. Small states generally fear much bigger ones, but they fear some more than others. Thus, they look to the lesser threat for protection and must pay some price for that protection. Today the United States is blessed, in a sense, to be perceived as the lesser evil by many states that have to face neighbors such as Iran or Russia.

Power-wielders must also master personal skills and the art of manipulation. The experienced power artist almost always begins by establishing political support at home or at least the illusion thereof. Without that key domestic support, an adversary can compel him to fight battles at home rather than abroad. In such cases presidents have to convince their counterparts abroad that they, personally, possess sufficient will to overcome their lack of political support at home.

At times, artistry requires a reputation for steadiness and reliability, as the Truman team had in establishing its commitment to Europe's defense after 1945. At other times, artistry calls for just the opposite quality: unpredictability. President Nixon tried to convince Moscow, Beijing, and Hanoi during the Vietnam War that he could be crazy: that is, that he was capable of using nuclear weapons to avoid defeat. It didn't work, but it was an interesting approach.

At all times, the power artist must establish and maintain his credibility. He must convince others beyond a reasonable doubt that he will and can do what he says, i.e., that he will deliver on promised and hinted-at actions, even at considerable cost to himself and his country. George W. Bush used military force in part to impress upon others that since he had done it once, he could be depended upon to do it again and again. If he had won and sustained victories in Afghanistan and in Iraq, he would have amassed great power. All too predictably, however, he got bogged down in both places and thus seriously undermined his own credibility. Clinton ran into a similar problem with Serbia. The Serbs didn't believe that he would ever resort to force with regard to Bosnia or Kosovo, because he had sidestepped such action so many times before.

James A. Baker III, George H. W. Bush's secretary of state, nurtured his reputation as a mean son of a bitch who would do whatever necessary to get his way. That reputation was central to his success in convincing most of the newly former Soviet republics at the end of the Cold War that they had better return the nuclear weapons stored in their territories to Moscow—or the United States would not protect them against the Russian bear. Somehow or other, even Bill Clinton managed to achieve the same result in denuclearizing Ukraine and Kazakhstan.

And American leaders would do well to learn, finally, that power shrinks when it is wielded poorly. Failed or open-ended wars diminish power. Threats unfulfilled diminish power. Mistakes and continual changing of course also diminish power.

It's still almost irresistible to think of military force as power, as did Machiavelli and others of yesteryear and yesterday. Force certainly looks like power in terms of getting others to do what they don't want to do. But war itself is essentially a physical act, whereas power is essentially psychological and political.

Of course, there are places where the two overlap, but the distinction is very much worth preserving. It's important for leaders to keep in mind that they should go to costly and uncertain wars over vital matters only after their power has failed. They, and particularly American leaders, need to focus on exhausting all plausible actions before pulling the trigger. The line between force and power is a thin one and should keep leaders alert to searching for creative means of getting what they want by pressure and coercion, short of the great financial and human sacrifices required by war. War represents an order of costs and consequences entirely different from any imaginable exercise of power or psychological arm wrestling. If leaders lose sight of the line, they will slip far more easily across it and into the dangerous world of war and all its incalculables.

None of this is to say that force and power aren't closely tied in many ways. First, a formidable military capability—the product of expenditures, investments in technology, organization, leadership, and the like—can be used by leaders to send power signals with and without words. Second, once fighting begins, certain targets can be hit or not, and certain weapons fired or not, to convey clear messages about further risks and benefits to the target country. But leaders would be foolish to forget that in wartime such clever maneuvers and signals usually are more easily conceived by Dr. Strangeloves than executed by sensible mortals, and that controlling or terminating a war often proves elusive. Third, the threat of force is, of course, a mother lode of power. The aim of a threat is to convince adversaries that they must reconsider their behavior or face considerable pain. Threats are about generating worry—the emotion that is central to the successful application of the psychology of power. Fourth and finally, leaders can and do convert military victories into power. Conquests at reasonable costs serve as warnings to future enemies of what might happen to them.

A reputation for winning wars at acceptable costs remains one of the most potent psychological tools in the arsenal of power. But it has proved dangerous time and again for leaders to be carried away by the power of war. The psychology of power requires control and careful manipulation, and wars, by their very nature, often spiral out of control, with costs and consequences soaring beyond any reasonable expectations or forecasts. The single most challenging task, of course, is to weigh the costs and gains of going to war against continuing the more subtle and intricate maneuvers of power that may not offer high prospects of success.

THESE TRADITIONAL CALCULATIONS OF war and peace have become increasingly complicated by the exponential increase in the importance of international economics. The enormous growth in the frequency and volume of business and financial exchanges worldwide certainly qualifies as a revolution. But it is a revolution more in economics than in politics or strategy. And it has neither preempted nor totally transformed the rules of power.

Rather, the new importance of economics alters power calculations, making them more complicated, and paradoxically giving new leverage to the stronger as well as to the weaker. But it has to be kept in mind that the resulting mutual dependencies have not eliminated the risks of war, or even measurably reduced them. Tribes and nations have often done the most killing among those with whom they have long traded animal skins and widgets, as the Europeans taught humankind.

All this said, the new economic patterns have already begun to occupy the attention of those responsible for state power. Economic transactions such as aid, trade, and investment, as well as membership in international economic institutions, exist for mutual economic and political benefits. But when governments conduct these activities, they more often than not intend to coerce, pressure, reward, or establish conditions for later coercion and pressure. Governments can give or withhold these opportunities or offer better or worse terms, and thereby incorporate economic transactions into the power relationship. And of course, like military capability, a strong and vibrant economy is, in itself, a source of power.

Oil-producing countries use supply and price controls as power tools, but mainly to affect economic rather than foreign policy matters. When major powers like the United States institute economic sanctions to deny a country a variety of economic dealings, however, it is almost always for foreign policy and strategic ends. Even the so-called humanitarian aid provided to poor nations by the United States and other countries does double duty as a means to bolster the donors' leverage on other matters. More than ever, economics has become ingrained in power relationships, and leaders are just beginning to sort out how it can be made to serve larger strategic and power ends.

POLICY AND DIPLOMACY ARE the words and deeds by which most power is conveyed. They are the means to signal what's at stake and to present incentives and disincentives; and when well conceived, they lay out a path for resolving difference. For example, the United States and like-minded big powers have been presenting Iran with an array of benefits it can enjoy in exchange for abandoning its nuclear program, as well as the penalties it will pay for retaining that program. So far, that path has proved neither sufficiently attractive nor adequately frightening. Some think the policy needs a broadening of carrots and sticks; others believe only sticks will work. Similarly, Washington is suggesting to Arab countries what they will have to do if they want the United States to pressure Israel. The path calls for them to recognize the state of Israel and pledge serious amounts of aid to the Palestinians.

Policy and diplomacy are not by themselves power. But well considered, they can assist the application of power—just as bad policy can undermine it. High-level policy pronouncements are the first place other countries look to gauge how Washington will wield its power.

WHEREAS CONSERVATIVES LOSE TRACK of power's meaning when they hearken only to the sound of the cannon, liberals levitate above reality when guided by their hearts. Liberal levitation doesn't occur in most societies, or particularly among the leaders of most countries. It's been mostly an Anglo-American phenomenon, and in America the heart has led to confusing power with its cousins—ideas, values, culture, leadership, and persuasion.

To forestall heart attacks among my liberal readers, I readily concede that I, too, love ideas, values, culture, leadership, and persuasion. But while these can foster or harm the application of power, they do not represent power in the international realm, except under the most unusual circumstances. Leaders of other nations, bound as they always are by their own interests and politics, are not going to do something they don't want to do because they've come to admire our values or because American leaders have succeeded in making previously unconsidered intellectual arguments of overwhelming and dazzling elegance.

There were moments after World War II when American ideals did possess a kind of power of attraction. New states emerging from the colonial yoke admired the Declaration of Independence, the success of America's economy and weaponry, and American anticolonial policy. Many were drawn to Washington's side in those years. However, it was undeniable that most of the new states remained neutral between a communist totalitarian Soviet Union and a free democratic America. As the Cold War dragged on, the United States lost further ideological ground because of its support for dictatorial regimes in the fight against communism, and especially because of the Vietnam War. There was a moment of rejuvenation when American ideals clearly played a part in the revolt of the Eastern European states against the dying Soviet Union. But it is well to remember that these upheavals had more to do with the collapse of Soviet power, and the poverty of the East compared with the West, than with the attractiveness of American ideals.

There may have been another triumphal moment when America stood as the sole remaining superpower, a position presumably signifying the superiority of its ideals, values, society, and government—a victory for freedom and capitalism. But that superiority vanished quickly when many around the world began to equate America's championing of globalization with a scheme to advance American greed. The Iraq War further tarnished international admiration for American ideals. As a result, there exists today an enormous divide between the values of the United States and those of many other parts of the world, reminiscent of some of the worst religious schisms in history.

I actually believe that America does have good ideals, far nobler than those found in most countries. But like any great power, America has worldwide responsibilities that conflict with its ideals. A Chinese diplomat recently berated me about just such U.S. double standards. I replied: "We don't have double standards; we have quadruple standards. It's the fate of every great power, as you yourselves must see."

Like ideas, values don't travel well and are therefore difficult for governments to translate into power. In the first place, it's not as if

America possesses a national consensus on values. Whose values are to be exported—those of Henry Kissinger, Bush Sr. or Jr., Jimmy Carter, or Bill Clinton? Americans may once have had widely shared values, but that golden era is long gone. While almost all Americans can agree on the fundamental virtues of freedom, equality, tolerance, and democracy, they differ wildly among themselves about whether and how to foster these values abroad—by force of arms or by example, by staging quick elections or by long-term political evolution. Second, even if Americans agreed, other societies place different weights on the same values. Besides, most of those societies attach greater importance to economic security than to political values. A chasm divides the American ideal of separating church and state from the Muslim ideal of theocracy. The same holds for the role of women in society.

It is even more far-fetched to treat culture as a source of power, let alone a means of exercising power. Granted, there are enormous audiences abroad for American music, television shows, and movies, and hordes buying American clothes and perhaps even literature. But the U.S. government does not wield music and movies. And although Americans pride themselves on their multiculturalism and tolerance, many others around the world don't share these attitudes. Perhaps after foreigners study in America and return home, they carry with them some understanding of us and our culture. Student exchange programs based on this hope deserve our support.

Let's gladly pocket and prize whatever goodwill comes from such attractions, but we should also avoid the delusion that the United States is wielding power in the process.

A similar illusion holds for leadership in the international arena. Leadership—derived from personality, charisma, message, political strength, and circumstances—can often be critical in mobilizing public opinion within nations. Also within nations, leadership succeeds on occasion in inspiring citizens to do what they would otherwise not do. Franklin D. Roosevelt and Reagan effectively orchestrated public support. Legislators and even the press fear to oppose a popular president, so leadership at home can produce power abroad.

But leadership on the international plane is quite another matter, and it is difficult to find examples of nations making concessions they otherwise would not have made purely because of another nation's leadership. Woodrow Wilson may have been a minor exception. He ignited a fever among Europeans for the League of Nations, for which European leaders initially had little enthusiasm. President Kennedy possessed great personal charm and popularity, and he was certainly admired in many other countries. But it is hard to document how such emotional responses led to concessions from other governments. It is extremely difficult for even a great leader of one country to out-lead a leader of another country, or to out-duel him in the politics of his own country. Why should foreign leaders be swept off their feet by American presidents' charisma any more than American presidents are carried away by the charm of foreign leaders?

Probably the biggest challenge to thinking about power as psychological pressure and coercion comes from the advocates of power as persuasion. But just as power is not force, neither is it persuasion. If power's roots are psychological, and if force is essentially physical, persuasion is basically an intellectual and emotional process. It amounts to altering another's thinking with arguments, facts, and appeals to feelings such as compassion or empathy.

Persuasion, or what I call convincing others that you understand their interests better than they do, sometimes works in personal relations, and occasionally in domestic politics, but only very rarely in the international arena. Leaders know their countries' vital interests. This knowledge is typically forged in the fire of domestic politics. And leaders also know that there's a high price to pay when they compromise on those vital interests.

Americans are forever trying to persuade others to stop killing one another and to enjoy the economic benefits and security of peace. After several centuries, that argument certainly prevailed in Northern Ireland. There are countless examples, however, of persuasion and good sense not being persuasive. Bosnia, Kosovo, Somalia, Kashmir, Darfur, and Rwanda come to mind. These peoples distrust and hate each other, and distrust and hatred trump persuasion every time. In fact, it's hard to find examples in international affairs where persuasion trumped a leader's concerns over his core power and political survival.

Even in universities, the so-called citadels of reason, it's difficult to discover examples of professors changing their minds, not to mention one another's minds, because of reasoned debate. It is not much easier with political leaders. That said, it has become a political necessity to try persuasion and reason before power and pressure, and this is fine and good. It can be helpful in building domestic and international support for future stronger actions.

THERE IS A TIME and place for the exercise of the entire spectrum of power and power-related actions, ranging from trying persuasion to offering incentives to resorting to the use of force. But the results of attempting all three are usually unsatisfactory. Generally, if anything is to be accomplished in the international arena, if problems are to be solved, it requires the effective employment of power. And power itself, as we have seen, is increasingly difficult to wield successfully. That's why it's more important than ever to understand power correctly. Every American administration since 1945 has struggled over how to think about and wield it, and the battle lines have never been more clearly drawn than now.

CHAPTER 3

POWER IN THE AMERICAN CENTURY

Americans have seldom stopped arguing about how to wield power in a nasty world. From Hamilton and Jefferson through Acheson and Kennan to Krauthammer and Nye, the disputes persist— unabated, unresolved, and too often unilluminating.

Two prototypes of foreign policy have always throbbed in American hearts, sometimes in the very same heart— the tough-minded realist, Alexander Hamilton; and the philosophical idealist, Thomas Jefferson. They fathered what would become a foreign policy civil war in the United States, notable for its ideological rigidity, political savagery, and historical longevity. Battles between their followers have inspired schizophrenic foreign policies and both triumphs and tragedies abroad, as the country has struggled to decide how to frame American objectives in the world and how to think about American power.

Their ideas have lived on in countless variations for more than two centuries. President Truman's brilliant advisers, Dean Acheson and George Kennan, both sprang from a Hamiltonian mold. They were professionals, not given to idealistic passions, but they were surrounded by fanatical Hamiltonian and Jeffersonian true believers. President Eisenhower was a Hamiltonian, although his secretary of state John Foster Dulles seemed to be both an ardent Jeffersonian and a determined Hamiltonian not unlike President George W. Bush five decades later. The presidents in between had their own intriguing variations.

For the last two decades at least, the Jeffersonian and Hamiltonian progeny have been barely recognizable. The Jeffersonians now appear in the tenuous union of liberals and moderates, while the Hamiltonians congregate in the philosophically strange alliance of conservatives and neoconservatives. But the influence of the two original debaters can still be detected in the advocates of soft power—the use of persuasion and values—and the hard-power crowd who champion the use of force and the application of America's will and staying power.

HAMILTON, WASHINGTON'S TREASURY SECRETARY, and Jefferson, the nation's third president, could not have been farther apart in their approaches. The former pressed for strengthening America's economic and military base for the inevitable confrontations with the world, in order to protect U.S. interests. The latter advocated a U.S. policy designed to promote the American ideals of freedom and democracy. Truman's secretary of state, Dean Acheson, argued for negotiating from strength and more or less shaping the terms of agreement—or not negotiating at all. George F. Kennan, his chief policy planner, was more inclined to try to resolve differences through negotiation and believed that the fate of nations turned mainly on their domestic successes or failures. The columnist Charles Krauthammer today maintains that power is best wielded by the use of force and by an unswerving will to stand up for American interests. Professor Joseph Nye of Harvard sees the world as moved far more by the example of America's values, its powers of persuasion and the gentle application of its leadership.

These thinkers and those who agree with them have rarely been inclined to compromise with one another. The combat among them was and remains a game of winner take all. This accounts for the all-too-frequent swings in U.S. foreign policy and for the continuing uncertainty and confusion over how to use American power.

HAMILTON FREQUENTLY LECTURED HIS contemporaries on the blessings of pursuing the classic path to protecting national interests: first, make the United States a thriving economy, and then turn it into a major military power. There was no other way, Hamilton said, to safeguard American interests in a selfish and nasty world. A few Americans eager to construct a huge continental nation grasped Hamilton's fundamentals of power politics, although the majority of his contemporaries preferred the lofty ideals of Jefferson. The world would follow America, Jefferson believed, if Americans followed their nation's founding ideals of freedom and equality. Americans shied away from Hamiltonian words such as "self-interest" and "power." They smacked of being, well, too European for New World tastes. The language and underlying thoughts sounded too similar to Old World thinkers, and to the Continent's penchant for perpetual wars, which most Americans sought to escape. Theodore Roosevelt carried the Hamiltonian banner into the twentieth century, albeit sometimes expressed in Jeffersonian rhetoric. But he was known abroad for building the "Great White Fleet" to announce the United States as a great power. Woodrow Wilson's idea of making the world safe for democracy and resolving conflicts through the League of Nations suited America's self-image, although not its isolationist politics.

After World War II, the United States could no longer escape the political struggles of Europe and the Soviet Union. The debate over our foreign policy taxed the boundaries of the political thesaurus with such barbs as "commie," "pinko traitor," "soft on communism," "warmonger," "appeaser," and "cut and runner."

There was something about the subject of national security that touched a central American nerve. It went beyond survival to questions about who we were and what we aspired to be. National security was more than a policy; it was a passion that brought forth the nation's best and its worst.

WHILE PUBLIC DISCOURSE ON American foreign policy often ran to the vacuous, the vicious, and sometimes even the libelous, there were moments when the debate—inside the government and among a small but steadily growing circle of foreign-policy experts—shone with America's pragmatic brilliance, when some of the nation's finest thinkers made some of the nation's greatest foreign policy arguments. Take, for example, Dean Acheson and George Kennan, the co-architects of America's brilliant postwar strategy to contain the Soviet Union and make the world better and safer through such formidable institutions as the Marshall Plan, NATO (Kennan waffled here), the World Bank, and the General Agreement on Tariffs and Trade.

If they had been characters in a Hollywood movie, the difference in their appearance would have signaled their differences of opinion: Acheson looked every inch the swashbuckling British brigadier complete with a theatrical upturned mustache that set off his perfectly tailored double-breasted suits, and added to his commanding presence. Kennan also favored traditional attire, but of an unflamboyant type that underscored his low-key demeanor and the grace and precision of his writing. Acheson, the brilliant lawyer, was always prepared for intellectual combat; Kennan, the brilliant diplomat, never avoided it.

These two giants framed a sophisticated policy debate, at least within government circles. Both adjusted their views over time and with changing circumstances, as befits serious thinkers. For the most part, they agreed on the goals of halting the expansion of Soviet power and eventually bringing down the Soviet regime, reconstructing postwar Europe and Japan, and expanding American power. They disagreed on the means or the power needed to achieve these noble ends.

Kennan gave the basic American policy of containment a new depth in a much admired long telegram from his diplomatic post in Moscow in 1946. It was published anonymously in the journal

Foreign Affairs, and came be to known as the "X" article because that was the entire byline. He wrote that the Soviets would try to extend their dominion, but would back off if the United States met these advances with counterpressure. They would not push conflicts to the point of confrontation, he believed, both because they recognized American strength and because their overriding aim was to retain power at home. In Kennan's view, the Soviet leaders would not risk their internal positions for external gain. If Washington applied the necessary, though not excessive, counterpressure, he predicted, Moscow would eventually collapse from within, the inevitable victim of its own corruption, inefficiency, brutality, internal contradictions, and fundamental flaws. Acheson rarely speculated on whether the Soviet empire would implode or how. Perhaps he felt that war between the superpowers was inevitable, but he didn't want to say so.

Both Acheson and Kennan believed in the centrality of economic power—interestingly, the former more than the latter. Both saw that economic well-being would make countries less vulnerable to communism and to pressures from the Soviets. Both regarded America's economic largesse as one part charity and nine parts hardheaded investment in American national security. From there on, however, they fought: Acheson from the heights of the establishment and Kennan from the groves of academe.

While Kennan argued that Moscow was "impervious to logic of reason," he went on to stress, "There is no reason, in theory, why it should not be possible for us to contain the Russians indefinitely by confronting them firmly and politely with superior strength at every turn." He called for "the manipulation of our political and military forces in such a way that the Russians will always be confronted with military strength." Acheson certainly found little to dispute in the plain meaning of these words.

But Kennan contended he had been misunderstood, and by 1949, he publicly excoriated the Truman administration for being excessively confrontational and needlessly exacerbating tensions with Moscow. He argued that if the United States would negotiate with the Soviets to reduce mutual tensions and hostility, the Soviets would reciprocate. "The dog," he argued, referring to Moscow, "for the moment, shows no signs of aggressiveness. The best thing for us to do is surely to try to establish, as between the two of us, the assumption that teeth have nothing to do with our mutual relationship." Kennan wanted to focus instead on protecting certain vital areas, such as Western Europe, and in that way to keep the Soviets boxed into their own sphere of influence in Eastern Europe.

While neither man was dogmatic about responding to Soviet pressures, Kennan preferred diplomatic, economic, and psychological power in contrast to Acheson's heavy emphasis on military and economic power. In 1949, speaking at the Council on Foreign Relations, Kennan cautioned, "We must refrain as much as possible from making the present East-West line a hard and fast one and should continually engage in negotiations with the Russians, even though we must recognize that they will consume needless time and that we cannot hope for success in terms of years." A year earlier, he had gone so far as to argue that the Western powers should withdraw their occupation troops from Germany, thus unifying Germany, the center of the Cold War, averting the division of Europe, and thereby lessening tensions with the Kremlin. Fortunately, he would never resume this train of thought. Nor did he hesitate in 1950 to endorse U.S. military force in Korea, and when China's entry signaled Soviet involvement, he fully backed Acheson's strong response.

Although Acheson initially favored engagement with the Soviets, even backing a plan to share nuclear technology with them, he soon began moving in the opposite direction. He saw negotiations as increasingly useless and naive. He declared: "You cannot sit down with [the Soviets]." He came to believe that the key to dealing with Moscow was establishing "situations of strength" throughout the world. "You can dam [communist ideology] up, you can put it to useful purposes, you can defeat it, but you can't argue with it."

By the time Acheson became secretary of state in 1949, his emphasis on military strength over diplomacy had become official American policy. He and his allies solidified their victory in a famous 1950

National Security Council study that came to be known as NSC-68. This seminal document saw a growing Soviet threat to U.S. interests worldwide. America, the authors argued, should counter this threat by vastly increasing its military capabilities—both conventional and nuclear—and by establishing "situations of strength" around the world. Once all this was in place, Moscow would have to "recognize facts."

Years later, they battled one last time, over Vietnam. Kennan opposed American involvement from the outset. He correctly saw that American leaders didn't know a thing about Vietnam, about its culture and colonial past, or, in particular, about the enormous power of Vietnamese nationalism. To him, Vietnam was essentially a diversion from the main theater of reckoning in Europe. The United States could only get mired down in Indochina and lose, he concluded. Early on, Acheson saw Vietnam less as a living country and more as a vital square on the strategic chessboard, where China and Russia were testing American power and will, and where Washington would have to prevail or suffer the strategic consequences. But by 1968, Acheson had come to believe that the war had been lost, shifted his ground dramatically, and urged President Lyndon B. Johnson to cut his losses.

The Kennan-Acheson disputes enriched the internal debates, and the policies produced by them held up well throughout the Cold War. Acheson's resistance to negotiations with Moscow was spot-on in the early years of the Cold War—when Western Europe's collapse and Moscow's intransigence would have led only to unilateral U.S. concessions. At that point, the focus had to be on shoring up friends. Acheson also correctly pressed for the restoration of U.S. military strength. But Kennan was correct, as well, in stressing the Soviets' caution and aversion to high risks, as long as they knew Washington wasn't a pushover. Most important, he was right to sound alarms about interventions in the Third World that would inevitably divert and weaken American power. He was also right to insist on the role of diplomacy, particularly as Soviet military might grew along with Soviet assertiveness. Finally and notably, these two policy giants agreed that American power was best exercised through multilateral institutions such as the Marshall Plan and the World Bank, in which Washington could use its power to lead while minimizing resistance from followers.

THE SOPHISTICATED POLICY EXCHANGES between Acheson and Kennan gave way in the Eisenhower administration to a schizophrenic period, with rhetorical excesses by Secretary of State John Foster Dulles, on the one hand, and very cautious actions, by the president, on the other.

Dulles was the public face of the administration and expressed the government's most confrontational rhetoric—advocating a "rollback" of Soviet power from Eastern Europe. Dulles couldn't bear the language of Truman's containment doctrine, which, to him, implied acceptance of Moscow's presence in the heart of Europe. Eisenhower didn't like Dulles' rhetoric, nor did America's allies, but Ike indulged Dulles. Perhaps he reckoned that this rhetoric calmed the nerves of the strident right-wing Republicans. But Eisenhower kept his distance from Dulles' insistent pressure to challenge Moscow in Europe. He even refused to aid the Hungarian revolutionaries in 1956 for fear of sparking a wider war.

Eisenhower didn't like war and didn't like communists. His vast military experience made him properly reluctant about using military force, save for clear-cut threats. His experience with Moscow made him appropriately skeptical about trying to compromise with its extravagant demands. So, he rarely drew the sword and rarely parlayed. He didn't venture into summitry until the very end of his second term.

Only once did Eisenhower assert American power against the Soviets and Chinese, and that was at the beginning of his administration. He was determined to end the Korean War and let the communists know he was ready to use the American nuclear arsenal. The United States had the nuclear edge, and Ike played that advantage. But he tied his threats to cease-fire compromises as well. His only other outright assertion of power was against America's own allies, France and the United Kingdom, to halt their invasion of the Suez Canal in 1956.

Eisenhower based his national security policy on three strategic elements. The first was covert action against communist-leaning governments in the Third World, most notably in Iran in 1953 and Guatemala in 1954. Second, in what was called the Eisenhower Doctrine, he asserted his intentions to protect U.S. interests in the Middle East with economic aid and military training. He did not press for permanent bases and troop deployments. Third, he developed a defense policy known as the "New Look." It featured modest expenditures and deterrence of wars rather than massive arms build-ups with which to wage them.

Eisenhower was against militarizing U.S. foreign policy. Whenever he could, he even went so far as to play down threats, as he did when Americans panicked over the Soviet launching of Sputnik in 1957. Instead of using this stunning advance to justify major increases in military spending, he called upon Americans to spend more on math and science education. Nor was he about to squander American power in Asian land wars. He replaced French troops in Indochina with U.S. military advisers, but not with U.S. combat troops. Eisenhower's hallmark was to warn against both anticommunist hysteria and excessive Pentagon budgets, as he did in his farewell address warning his fellow citizens against the undue influence of the military-industrial complex. To Eisenhower, American power was to be carefully husbanded and applied only sparingly, when necessary.

PRESIDENT KENNEDY CHARGED THAT Ike's husbanding of American power had led to a loss of American prestige and power in the world. He argued that the Soviet threat was on the rise, and that Eisenhower had done little to thwart it. He promised to get America "moving again" at home and abroad. During his brief tenure, he experimented with almost every instrument of power, getting very mixed results, and seemed to find a comfort zone only after the Cuban missile crisis of 1962.

At the outset, he was bent on restoring American power with a mix of big defense spending increases and high-stakes summitry with the Soviet premier Nikita Khrushchev. He quickly traveled to Vienna for a summit conference with Khrushchev, and, just as quickly, regretted it. He accurately sensed that Khrushchev viewed him as an inexperienced pushover. As for his desire to demonstrate U.S. will and power, that soon morphed into the fiasco of the 1961 Bay of Pigs invasion in Cuba.

But Kennedy, ever sensitive to the effects of failure on power, redeemed himself a year later, in 1962, with his management of the Cuban missile crisis, the single most brilliant display of American military and diplomatic power in Cold War history. He masterfully assembled naval and air superiority around Cuba and backed it up with a combination of tough public and flexible private diplomacy. He followed this triumph by initiating bold negotiations with Moscow to ban nuclear testing aboveground.

For all his newfound prestige and confidence, however, Kennedy could not come to terms with the deteriorating situation in Indochina—erratically responding by making deep public commitments to Vietnam, then saying that salvation was up to the Vietnamese, then increasing U.S. forces there, and then pretending to reduce them. At the moment of his assassination, America's standing globally was soaring, demonstrating the dynamic president's great skill at appealing to peoples around the world, an appeal he did not live long enough to exploit. He also set military doctrine on a more credible course by diminishing Eisenhower's reliance on nuclear might and initiating the buildup of U.S. conventional (nonnuclear) forces. All this put the United States in a better position both to threaten and to fight.

KENNEDY COULD GET AWAY with sending mixed signals on Vietnam because the situation there hadn't reached a point of desperation where the only option was to escalate the war or face the humiliation of defeat. President Johnson didn't have that luxury as communist gains threatened victory. And while Kennedy was busy enhancing America's military might, it was already in place and ready for use by LBJ. By 1965, the president who dreamed of making his mark at home with his "Great Society" programs found himself ensnared in a war with no feasible escape routes—and with virtually all of his country's power focused on that one tiny spot in Asia.

I was director of policy planning and arms control in the Office of the Secretary of Defense during Johnson's last two years in office, 1967–1969. Having just turned thirty, I was part of the cadre of inexperienced national security experts drafted into the Pentagon by Defense Secretary Robert S. McNamara, the chief architect of American involvement in the Vietnam War. I have never felt so much tension and pressure in any position in my life: America was on the path toward deploying more than 500,000 servicemen and -women in the theater of war, with deaths ultimately to exceed 50,000, countless wounded, and the nation sharply divided over our involvement in Southeast Asia. A few of us civilians in the Pentagon were among the few in the national security bureaucracy who were arguing for de-Americanizing the war and starting negotiations. It was hardly a dovish stance, but in the heated atmosphere of Washington's anticommunists and antiappeasers, we found ourselves often targeted and always vulnerable.

In the meantime, Johnson kept increasing U.S. military involvement, putting almost all of America's power and prestige into Vietnam; little remained for anything else. He cashed in most of his chits with friends and allies and called in all favors in return for their assistance in Vietnam. The United States had been protecting many of them, and it was payback time, although not much was paid back. Johnson was obsessed by the fear of losing this war, and he had little time or power left over for other world affairs.

Vietnam was the culmination of the decades of tension between the United States and its allies over the former colonial world. Once shorn of their former colonies, our Western European allies wanted little or no part of the high-profile military operations there. At the same time, the allies were becoming increasingly averse to risky confrontations with Moscow in Europe. America's allies developed an intensely conflicted attitude toward American power—at once taking it for granted and relying on it, while also worrying that it would get them into trouble with their overbearing neighbor to the east, the Soviet Union. By the end of Johnson's tenure, American power was drowning in Vietnam.

THE GENIUS OF PRESIDENT Richard M. Nixon and Henry A. Kissinger, his principal adviser, was to let the victim drown slowly while they steered the world's attention in another direction—to the most dazzling and theatrical display of American power ever. Whether by design or not, they dragged out the Vietnam War, perhaps hoping for victory, but not expecting one, and made their main focus the ushering in of the most active and wide-ranging period of high-wire, high-stakes diplomacy in American history. For this, the president and the nation paid the price of the Watergate scandal and of profound domestic social upheavals.

The two master strategists arrived in the White House with the view that U.S. power was in decline, and that Americans had lost their resolve to remain a great power. To overcome these weaknesses, they designed a theater of diplomacy, a nonstop show with bells and whistles: They calculated that if their diplomacy dazzled the world and succeeded, they would restore America's lost power.

Nixon and Kissinger certainly accomplished a great deal—and certainly knew how to exaggerate their extraordinary accomplishments. They overstated American weakness and war fatigue to make their struggle to overcome them appear all the more heroic. More credit to them, they also knew how to make their diplomatic triumphs look like Herculean feats, although these feats still left key problems unresolved.

Again, to their credit, they were masters at using a combination of mirrors and fears creatively, at least outside Indochina. To name but three of their greatest diplomatic triumphs: the dramatic opening of relations with China, the broad-ranging and daring arms control talks with Moscow, and a masterpiece of a settlement between Egypt and Israel after Israel's victory in the 1973 Yom Kippur War. The opening in Beijing gave the United States new leverage over both Moscow and China, and Nixon and Kissinger's wheeling and dealing in the volatile Middle East established America as the sole peace negotiator acceptable to all regional parties.

In effect, the Nixon-Kissinger team mapped out a new kind of international power base for the United States—one rooted in America's unique position as the only nation most adversaries would work with. The strategy gave the United States a new aura of indispensability and, in turn, cushioned the defeat when it ultimately came in Vietnam in 1975. Almost every foreign policy expert (especially including Nixon and Kissinger) had predicted that Saigon's fall would trigger a strategic disaster for Washington, precipitating a series of events that would inevitably lead to America's allies falling like dominoes to Soviet communism. But no country in Asia wanted to see a dispirited and depleted America, because Asia clearly grasped the centrality of a strong United States to the region's security and economic future.

Though Nixon and Kissinger had left American foreign policy and power in reasonably good shape, Jimmy Carter did not see it that way. To him, Nixon had exaggerated the importance of Soviet-American relations and had undervalued North-South ties—relations with the developing world. Perhaps most tellingly, Carter believed Nixon and Kissinger had utterly failed to play America's strongest power card: its values and its role as preeminent champion of human rights.

That said, however, Carter was almost always of two minds about U.S. power: one, the hard-edged, Soviet-leery approach of Zbigniew Brzezinski, his Polish-born national security adviser; the other, the idealistic and lawyerly approach of Cyrus Vance, his secretary of state. He chose one, then the other, but rarely integrated the two into a coherent overall policy. The man who would become a very outspoken former president left everyone confused about if and how he would exercise American power.

I worked for Vance as the assistant secretary of state for politico-military affairs, responsible for Soviet-American arms control talks, conventional arms sales, and all matters relating to the use of force. My loyalty to Vance never wavered, but my views fell much closer to Brzezinski's. At the State Department, I was given the responsibility to reduce arms sales, despite the fact that I had just published an article in *Foreign Policy* saying that such sales were a "major instrument" of U.S. policy and power. I now found myself pushing for arms control agreements with Moscow, even though I believed that arms control talks could not achieve much because neither side was ready to give up much. I favored the deployment of the infamous neutron bomb (the nuclear explosive that would destroy human life, but leave buildings intact), not because I thought it was needed militarily, but because the Soviets had just deployed new missiles and were trying to prevent our modernizing NATO's capabilities in any way. Moscow, I believed, couldn't be permitted such a veto over our military decisions. These negotiations continued with Moscow, as our allies took Carter less and less seriously. By 1980, I myself ended up voting for Ronald Reagan over Carter.

The main battle of the Carter years was that between the Columbia professor and the Wall Street lawyer. Though Brzezinski outdid Acheson in his passionate anti-Soviet feelings, and Vance went well beyond Kennan's belief in the power of persuasion, Brzezinski and Vance stood even farther apart on most issues than had the two Truman aides. But they did share Carter's commitment to promoting human rights, which became a centerpiece of Carter's foreign policy. It sprang from Vance's conviction that Nixon and Kissinger had strayed too far afield from basic American values, and that their diplomatic maneuvers lacked any semblance of ethical ballast. The top tier of the Carter team all thought that holding the human rights banner aloft would restore America's moral leadership internationally, giving Washington renewed influence almost everywhere as people associated America once again with hopes for a better and more democratic life. But while the human rights campaign certainly did reestablish America's moral credentials, it also tended to undercut Carter's quest for agreements with the Soviet leadership, and it worried many of America's traditional allies, such as Iran and Saudi Arabia, whose authoritarian regimes felt threatened by Carter's high-minded rhetoric.

To further complicate the situation, the Vance-Brzezinski split was magnified by their supporters within the administration and even more so by the new political alignments on foreign policy in the country itself. Democrats had moved farther to the left while Republicans were moving farther to the right. Vietnam had been the trigger for this change. Most Democrats strongly opposed the Vietnam policy in the Johnson and Nixon years, and most Republicans backed it. But in a pivotal shift, outraged conservative Democrats—later to be known as neoconservatives—fled to the GOP, where they found hard-line soul mates. These migrations homogenized the formerly mixed political parties and reduced the overlap that had allowed for some degree of bipartisanship. Political room for maneuvering and compromises sharply narrowed, and the modern era of 24/7 partisanship was born.

In effect, the common ground of the Kennan-Acheson era had disappeared, and the two sides now clashed sharply over everything. Conservatives saw the Soviets as stronger and more dangerous than ever. Liberals, while decrying conservative exaggerations, demanded more arms control. Vance pressed for more mutual compromises with Moscow. Brzezinski, who didn't oppose negotiations with Moscow, soured on compromises, and believed that Moscow would give in if Washington stood more firmly. If this didn't work and the two sides failed to reach agreement, so be it.

Carter leaned to one, then to the other, and mostly toward Vance—until the Soviets invaded Afghanistan and the Iranians took fifty-two Americans hostage in Tehran. Those two blows changed Carter from a man who had initiated one of the most ambitious set of arms control negotiations ever with Moscow to a president who didn't want to negotiate with Moscow at all. He also announced the Carter Doctrine, which stated that the United States would not allow any power to control the Persian Gulf.

Still, Carter could not decide what kind of power to use in these situations. In Afghanistan, he took a tough stand and stepped up the supply of arms to the Muslim rebels fighting the Soviets. But in Iran, he neither took nor threatened military action against the Iranian hostage-takers, his failed rescue effort notwithstanding. Carter's four years of lurching between Brzezinski's distrust of the Soviets and Vance's unwavering support for arms control talks foreshadowed the new style in U.S. foreign policy—where presidents swung from hard to soft lines, and soft to hard, from threats to concessions and back, quickly and quixotically.

MUCH AS KENNEDY HAD felt impelled to make up ground lost by Eisenhower's restraint in world affairs, Ronald Reagan entered the White House convinced that Carter's vacillations had only encouraged the Soviets to expand their military power and frontally challenge American interests. Immediately, he ordered major increases in Pentagon spending, exceeding those of JFK, and hardened all U.S. negotiating positions with the Soviets. Reagan was putting himself in a position to make maximum demands, and if the Soviets refused those demands, he seemed fully prepared to live without arms agreements altogether. The Cold War, he believed, was reaching a crescendo, and his aim was to maximize U.S. power to pursue maximum goals against Moscow, leaving little time or resources for anything else.

Reagan punctuated his military buildup with rhetorical rockets, as when he called the communist superpower "an evil empire." Evil it was, but no president had ever called this spade a spade. This drove our European allies, always easily rattled, into fretting that Reagan was a greater threat to world peace than the Soviets. Allied unity hung in the balance until Reagan improved ties with Moscow late in his administration.

Reagan's devotees have lobbied historians to portray the Soviet collapse after he left office as the fruit of his hard-line strategy: that is, the strategy of driving up U.S. arms to drive up Soviet spending on arms in a weakened economy to drive Moscow into economic bankruptcy, and finally into total political collapse. But the weight of the evidence shows that the Soviets didn't try to match Reagan's military increases, because they couldn't. Their economy was already sinking from decades of mismanagement and corruption, and their political clout was undermined by their defeat in Afghanistan. For the most part, then, the Soviet system rotted away from within, just as Kennan had predicted it would.

But Reagan's devotees usually ignore the other half of his strategy, once America's triumph in the Cold War became clear. Reagan worked to make the most far-reaching arms control agreements with his Soviet counterpart, Mikhail Gorbachev. Both leaders pushed aside their hard-line advisers, and Gorbachev met Reagan more than halfway. To understate the situation, many of Reagan's advisers were unnerved by the mutual cordiality of the two leaders.

As this historic Soviet-American drama was unfolding, Reagan made some serious mistakes in his use of American power in other places. He was fixated on the Soviet penetration of Central America and moved swiftly against Moscow's allies in Nicaragua and El Salvador. He spent $4 billion in El Salvador alone, on military and economic support during the civil war, and launched a major covert military program against the communist-sympathizing Sandinista government. After Congress prohibited further aid to the contra rebels in Nicaragua, Reagan approved a bizarre and illegal program to sell arms to the Iranians, who were allies of the terrorists holding Americans hostage in Lebanon, in return for cash, which Reagan then transferred to the Nicaraguan contras. The Reagan team would fight communism—even at the expense of selling arms to a country, Iran, linked to terrorism and thus establishing the dangerous precedent of trading arms for hostages. In another, earlier, misstep in 1982, Reagan dispatched a small contingent of marines to Lebanon. More than a year later, on October 23, 1983, suicide terrorists drove a bomb-laden truck into the U.S. Marine barracks in Beirut, killing 241 servicemen. Suddenly, the president saw Lebanon not as a peacekeeping operation, but as a big civil war, and wanted out, fast. Four months later, with no serious effort to hunt down the killers or to inflict retribution, and after having called Lebanon vital to U.S. interests, he redeployed the remaining marines. Reagan and his team didn't realize that Lebanon represented the first terrorist blow of a new era—and they let the terrorists get away with it.

He did, however, take the right and decisive action later, in 1986, when he launched an air strike directly against the Libyan strongman Muammar Qaddafi. It was in retaliation for a Libyan-sponsored terrorist attack against U.S. soldiers in Berlin. Qaddafi survived, but his daughter was killed. It's possible that this reminder of American power may have contributed to Qaddafi's later decision to fundamentally alter his policies.

Looking back, we can see that Reagan began somewhat like Kennedy: bombastic rhetoric and stepped-up military spending to check presumed Soviet superiority, followed by negative reactions from around the world, followed by a diplomatic offensive. Perhaps the former set up the latter, but that's for historians to debate. But don't scoff at the likelihood that Reagan knew what he was doing. I interviewed him several times for a piece in *The New York Times Magazine*, and he convinced me that he understood being tough gave him the power to compromise. Or as James Baker put it to me, "The president was, after all, a C-plus student at Eureka College, but I've never seen a better bargainer."

George H. W. Bush began where Reagan left off—and bested him with a well-laid plan for wielding America's newfound diplomatic power to end the Cold War in peace. He and his team performed brilliantly. With Soviet leaders barely holding on to empire and country, the United States seemed to have the power to dictate the terms of defeat. But had Bush attempted to do that, he would have badly overplayed his hand. Instead, he sought to shape the

terms of the peace. Specifically, he arranged for the withdrawal of
Soviet forces from the heart of Europe—without making Mikhail
Gorbachev look powerless. If Bush had tried to break the negotiat-
ing bank and make Gorbachev pay dearly, almost certainly the last
Soviet leader would have resisted.

Bush, James Baker, his secretary of state, and Brent Scowcroft,
his national security adviser, also brought their diplomatic power to
bear in order to consolidate on Russian soil the nuclear weapons that
had previously been stored throughout the former Soviet Union.
They also worked with the Soviets to bring about elections in Nicaragua, which led to the defeat of the Soviet allies there, the Sandinistas. And they kicked off a promising, albeit short-lived, peace
process in the Middle East, involving almost every player in the
region, as well as key outsiders. Bush, Baker, and Scowcroft realized
the United States had the power not to dictate, but to get things
done through diplomacy.

Their sense of achieving U.S. aims by limiting those aims was
most deftly displayed in their handling of the First Gulf War. To
drive Saddam Hussein's army from Kuwait, they took their time,
and they limited their goals, both to obtain the blessing of the UN
Security Council and to assemble an impressive coalition. Their decision was controversial then, as now. Whether their call was right
or wrong, they kept their vast coalition together, including Arabs,
and established the United States as the world's indispensable diplomatic power and peacemaker.

The Bush team fared less well in coming to grips with failed or
failing states such as the former Yugoslavia and Somalia. They didn't
see any clear-cut American interests in these countries. Neither the
reports of immense human suffering, the heartrending images of
refugees and poverty, nor earning a reputation for being uncaring
troubled them. Nor did they understand that these nasty civil wars
could become the breeding grounds for terrorists. Brent Scowcroft,
Bush's national security adviser, sensed that they were entering an
era of new challenges and called for "a new world order." But nothing came of the idea. Bush's team performed better in putting the

old world's problems to bed than in getting ahead of the new ones. They grasped how power had to be exercised in a more complicated world, but they didn't come to terms with the new problems that world faced.

BILL CLINTON IGNORED MANY of Bush's diplomatic endeavors until late in his tenure. My first encounter with Clinton came when he was governor of Arkansas and I was the op-ed page editor of the *Times*. I phoned him to discuss a piece on infrastructure, and he took my breath away with his knowledge and analysis. After almost an hour, I finally interrupted to say, "So, you'll do the piece for me?" "Yes," he responded, "but what do you want me to say?"

Clinton could impress any world leader with his knowledge and power of expression. But he seldom looked comfortable wielding power either at home or abroad. Perhaps he believed that with the end of the Cold War and America's being the only superpower, he could escape history's ugly game of power politics. The gods of optimism appeared to have sprinkled stardust on the playing field: the Internet seemed poised to trigger unprecedented quantities of information that would eventually lead to peoples everywhere achieving freedom and, ultimately, even wealth. Globalization seemed ready to guarantee democracy, peace, and prosperity to all. Accordingly, Clinton pivoted toward domestic issues to make the American economy and society his top priority. The motto of his election campaign had said it all: "It's the economy, stupid!" Foreign policy would be subordinated to domestic policy, and the military and economic power so prominently featured during the Cold War would take a backseat to understanding others, friendly persuasion, the spreading of American values and culture, and the pursuit of common economic interests. Dire threats to national security seemed gone with the wind. An overall strategy for U.S. power didn't appear necessary. As problems arose, Clinton could simply serve up the needed dollops of American power, à la carte.

If there was any consistent pattern to Clinton's efforts, it was to dodge international bullets and stay focused on both the world economy and the American economy. Following up on the policies of Bush, in 1992 he signed the North American Free Trade Agreement (NAFTA), which increased trade among the United States, Mexico, and Canada. He also lifted the Asia-Pacific Economic Cooperation to the summit level. In 1995, the United States joined the World Trade Organization (WTO), designed to reduce trade barriers and resolve trade disputes among most nations in the world. Both NAFTA and the WTO cost Clinton heavily among fellow Democrats who argued that he was disadvantaging American workers and jeopardizing American competitiveness in his rush into the new, globalized world. He pushed to expand trade with China and a rising Asia generally. He built up economic capital impressively, but never deployed it for strategic purposes.

When it came to dealing with failed or failing states, Clinton was the very model of Carteresque confusion and inconsistency. He had slammed Bush for being indifferent to humanitarian horrors in Somalia and Bosnia, only to follow suit by scrambling to avoid military intervention. In fact, he began by pushing for a cut in U.S. troops in Somalia and pressed the UN to assume responsibility there. But then, after some military reverses, he added U.S. Special Forces back into the mix and misguidedly expanded their mission to hunting down Somalian warlords. And then, after some special forces were ambushed, he withdrew all U.S. forces entirely.

After three years of excuses for doing nothing about the Serbian genocide in Bosnia, Clinton finally took effective military and diplomatic action. The limited and surgical use of force and the brilliant diplomacy by Richard Holbrooke were probably the proudest moments in Clinton's foreign policy. The president's subsequent handling of the civil war in Kosovo reinforced the sense of his being on top of the failed-state problems of the new era. But then there was his failure to act to stop the genocide in Rwanda, the most indelible foreign policy stain on Clinton's record.

Nor did Clinton get his bearings on combating rogue states and terrorism. He said Saddam was planning to assassinate former

President Bush, then dutifully hit Baghdad with a few cruise missiles for a day. In 1998, he bombed Iraq for four days after Saddam interfered with UN inspectors. These actions surely left Saddam—and perhaps Milosevic as well—convinced that Clinton was incapable of being serious about using real force. And when terrorists tried to blow up the World Trade Center in 1993, and later struck U.S. embassies in Africa, he did nothing at all for five years, until 1998, when he ordered small cruise missile attacks against Afghanistan and Sudan. These erratic, fragmentary, and feeble acts showed that Clinton had no comprehension of the nature and the magnitude of the growing global terrorist threat, let alone what to do about it.

Clinton did not hesitate, however, to seize upon opportunities for conflict resolution. He wisely allowed the Israelis and Palestinians to negotiate directly and secretly in Oslo and then played successful host to the signing of their accords on the White House lawn in 1993. He jumped into the Haiti situation in 1994, sufficiently to restore constitutional order there, but not to resolve the larger problems. In 1998, he played successful peacemaker in Northern Ireland.

But then, in the waning days of his presidency, he bizarrely undertook the most far-reaching negotiations with both the Israelis and Palestinians as well as with the North Koreans. Bizarre, because he had only weeks left to serve out his term and thus could not bring U.S. power to bear to conclude the agreements, let alone the power to bind his successor to them.

Clinton departed office more liked than feared. In fact, he was not feared at all. The one exception may have been Belgrade, the recipient of a well-deserved military punishment because of both Bosnia and Kosovo. People in Eastern Europe certainly admired him for including them under NATO's protection. Generally, leaders and people around the world did like Clinton and Clinton's America, but they seldom did as he wished or demanded. The positive feelings—as well as the strong American economy—never translated into strategic power.

· · · ·

IF CLINTON BELIEVED THAT the new millennium meant the United States could escape power politics, President George W. Bush thought it meant that the United States could dominate power politics. The former hoped he would not have to fight; the latter believed he could now fight and win. Whereas Clinton felt he should focus on the economic dimensions of international affairs and avoid force as much as possible, Bush quickly concluded that threatening and waging war would provide the answer to all America's prayers for security. Clinton reckoned he was inaugurating a new era of noncoercive power. Bush calculated that with the Soviet Union gone and America's standing as the sole superpower, he was the beneficiary of a new golden era of American military power.

Bush and Condoleezza Rice, his first national security adviser and later his secretary of state, believed that Clinton's major mistake had been to focus American power on small-fry issues such as Bosnia, Somalia, and Kosovo, while overlooking the new emerging threats from major rising powers such as Beijing and Moscow. In the thinking of Bush's team, these major powers were the only ones that could hurt the United States, but the team always added Saddam Hussein for good measure, in private. It would wait for the right time to reveal this little secret.

But the team set aside this carefully conceived and bizarre big power balance-of-power theory when al-Qaeda terrorists struck on September 11, 2001. From then on, all powers would be subordinated to the "war on terror." With tremendous support worldwide, Bush ordered U.S. forces into Afghanistan to take out al-Qaeda and remove its Taliban hosts in Kabul. But before completing that task, he linked 9/11 to Iraq, with almost no evidence, insisting that Saddam was close to achieving a nuclear weapons capability, and attacked Iraq. The U.S. forces rapidly toppled the Taliban and Saddam, and then found themselves enmeshed in ongoing insurgencies for which they were not prepared.

Bush had warned Iraq, along with Iran and North Korea, in his 2002 "axis of evil" State of the Union Address. They were all rogue states, hell-bent on developing nuclear weapons and on aiding terrorists, and, he asserted, Washington would stop them. Even after he invaded Iraq, Bush continued to threaten Tehran and Pyongyang, but the quagmire in Iraq took away the power of his threats. Iran and North Korea continued their nuclear programs, and Bush—his threats ignored—later modified his line to say that diplomacy had to be given a chance to solve these problems and that diplomacy would take time. He never withdrew his threats from the table, but Americans and others seemed more worried by them than the Iranians or the North Koreans.

Nor did Bush ever disavow the announced Pentagon doctrine that the United States would prepare itself to take preemptive military action against any state it judged to be a serious threat. The announcement of the doctrine in 2002 caused a firestorm. What disturbed many at home and abroad was not that the United States had such a capability or doctrine, but that the Bush administration chose to make it public—with all its tricky ambiguities and its in-your-face threats. Preemptive attacks without hard evidence of imminent threats to America's national security sounded like preventive war, wherever and whenever Washington felt threatened. The doctrine of preemptive action sent a notice of intent, and the "axis of evil" speech had clearly put the targets on notice.

Until the last year or so of his administration, Bush asserted the vast powers of commander in chief and exercised almost complete control over Congress through an obedient Republican majority. No one in America could or did deter Bush from conducting foreign policy as he wished.

Bush and Clinton failed for almost opposite reasons. Clinton had been liked and not feared, but not liked enough for others to do his bidding. Bush had been feared and not loved, but not feared enough for others to submit. Neither Clinton's nor Bush's side, however, would abandon its essential beliefs about power. The opposing bands played on—through the 2008 presidential elections.

. . . .

FOR THE FIRST SIX months of President Barack Obama's administration, it wasn't easy to grasp how the president and his team
think about power. To begin with, they were clearly sensitive to
stage-setting power and the need to make foreign policy rhetoric sympathetic to most ears around the world. And this helped
reduce irritation toward America. Obama also showed a welcome
disposition to negotiate with adversaries but gave little sign of the
strategy and power he would employ in these difficult gymnastics.
There has been very little White House deployment of carrots and
sticks that could be decisive on key issues such as a political settlement in Iraq, Afghanizing that war, or making the tough bargaining choices in dealing with North Korea and Iran.

Thus, the Obama team hasn't shed many of the old hard-soft
power disputes between the moderate-to-liberal camp associated
in many respects with Professor Joseph Nye of Harvard (who had
served in the Carter and Clinton administrations), and the neo-
conservative-to-conservative camp associated with Charles Krauthammer, a Pulitzer Prize–winning columnist for *The Washington
Post* and a psychiatrist by training who thrilled top Republicans with
his sharp intellect. Both began with the correct premise that the
United States now sits at the top of the power mountain alone, the
Rome of its day, and then they parted ways.

Nye started this latest round of the ongoing debate with his 2004
book, *Soft Power*. Nye has been a well-known moderate and pragma-
tist, but interestingly and inexplicably, he took a stance in this book
that was clearly in the liberal tradition. He asserted that there are
three ways to get what you want from others: "One is to threaten
them with sticks; the second is to pay them with carrots; the third
is to attract them or co-opt them, so that they want what you want.
If you can get others to be attracted, to want what you want, it costs
you much less in carrots and sticks." Soft power is "not force, not
money," he argued, neither carrots nor sticks, which he confined to
"hard power." Rather, it aims to "engender cooperation" by "an at-

traction to shared values, and the justness and duty of contributing to the achievement of those values."

At different points, Nye elaborated his thinking as follows: Soft power "co-opts people rather than coerces them." Later, "Soft power rests on the ability to shape the preferences of others." Then, "Credibility and legitimacy are what soft power is all about."

As Nye's thinking evolved, he added economics to the soft-power quiver, whether wielded as carrots or sticks. In other words, he introduced economic coercion through the back door, but still called it attraction. Then, he added the final touch to soft power—military power. "Military prowess and competence can sometimes create soft power. Dictators such as Hitler and Stalin cultivated myths of invincibility and inevitability to structure expectations and attract others to join their cause. As Osama bin Laden has said, 'People are attracted to a strong horse rather than a weak horse.'"

Soft power now seems to mean almost everything. It includes military prowess (presumably demonstrated by military action) and all kinds of economic transactions involving the giving or withholding of money for coercive purposes, as well as the old standbys—leadership, persuasion, values, and respect for international institutions and law. Perhaps the advocates of soft power had trouble finding noncoercive examples of getting other nations to do what they didn't want to do. In any event, what Nye and others finally put forward was more realistic and closer to moderate positions than where they had begun.

Most recently, Nye teamed with the conservative Richard Armitage as coauthors of "Smart Power," a report published by the Center for Strategic and International Studies in Washington, D.C. Smart power is, of course, neither hard nor soft, but a combination of the two. But their concept really is a mechanical combining rather than a genuine blending of the two ideas, and it clearly leans toward the soft-power side. "Legitimacy is central to soft power," Nye and Armitage wrote; they then added: "Legitimacy can also reduce opposition to—and the costs of—using hard power when the situation demands. Appealing to others' values, interests, and preferences can, in certain circumstances, replace the dependence on carrots and sticks."

Nye and Armitage are not exactly knee-jerk liberals throwing around their foreign policy credentials. They are pragmatists and realists, however unformed their new version of power. But Charles Krauthammer and many conservatives are prone to caricature anyone who disagrees even slightly with them. This doesn't elevate the public debate, but it does give them an advantage.

Krauthammer, of course, never wrote a book entitled *Hard Power*, but he made no bones about favoring military power, threats, and the national willpower to see through challenges to a victorious end, or about being opposed to bargaining with anyone he judged to be the devil. He argued that America was "the dominant power in the world," and that it was therefore "in a position to reshape norms, alter expectations, and create new realities. How? By unapologetic and implacable demonstrations of will." He labeled those who called for multilateral diplomacy as "hopelessly utopian." The United States was in the right and had the right to act unilaterally.

In case anyone missed the point, he stressed this: "What stability we do enjoy today [in the world] is owed to the overwhelming power and deterrent threat of the United States." And then to demonstrate his admiration of that power, he wrote: "Most Americans did not even know that our special forces could ride horseback, train a laser on a tank and see it pulverized by a bomber that might have come from Missouri. . . . [T]he power and resolve that America demonstrated in Afghanistan have already deeply impressed the world." Others, like Yemen, he noted, would henceforth behave, and their restraint would "com[e] not from love of America. It comes from deep fear and newfound respect." If America's European allies wouldn't go along with this, he dismissed them: "We will let them hold our coats, but not tie our hands."

As for dealing with tyrants like Saddam who were developing nuclear weapons, he flatly opposed diplomacy, fearing that only worthless agreements and cheating would result. To Krauthammer, the only safe course of action was regime change or the overthrowing of evil leaders. Krauthammer's message to Saddam, the Iranians, and the North Koreans was unambiguous: "You will be not only disarmed but dethroned." He did not have much, if any, faith in what he termed "classical deterrence"—military action alone promises safety. As for America having to go it alone, Krauthammer counseled against worry: "The new unilateralism defines American interests far beyond narrow self-defense. In particular, it identifies two other major interests, both global: extending the peace by maintaining democracy and preserving the peace by acting as balancer of last resort." Other countries "will cooperate with us, first, out of their own self-interest and second, out of the need and desire to cultivate good relations with the world's superpower. Warm and fuzzy feelings are a distant third."

One final thought about Nye and Krauthammer—the ultimate irony, in fact—should not escape unnoticed. Both supported most U.S. military interventions in the last two decades. In fact, Krauthammer did so fewer times than Nye because he opposed the humanitarian interventions in Bosnia and Somalia. While foreign policy experts usually battle each other over policy, they tend to follow presidents to war together.

Left and right have come together in another area as well in recent times—in the realm of promoting human rights, democracy, and freedom: what now passes for morality and ethics. Traditionally, liberals always occupied this high ground, but beginning with Reagan in the 1980s, conservatives also began to fly under these colors. Both were joined most enthusiastically by many neoconservatives during the Clinton years. Thus the twenty-first century bears witness to articles such as the one by the liberal Ivo Daalder and the neoconservative Robert Kagan, proclaiming the need for a coalition of democracies (Peru, South Korea, America, and South Africa, to name a few) to advance democratic values and interests. The antirealists of both stripes finally discovered a common home.

THE HAMILTON AND JEFFERSON camps, along with their soft- and hard-power descendants, are still alive and well and largely intact. They are smart, knowledgeable, and talented professionals, and practiced warriors. They also continue to reflect the two gut impulses of American foreign policy—the power of love and the love of power, soft and hard. Americans don't want their power raw; it has to be sautéed in the best of causes. But Americans do love to feel that their country is powerful, and that their leaders will excel at transforming the world into a better and safer place.

Neither group has gotten power right. And as long as these two traditions, with all their flaws, continue to dominate the public debate, America's enormous world power will remain hobbled. But there is some light in this tunnel.

The Clinton administration, its occasional successes notwithstanding, largely weakened the Jeffersonian soft-power school. And George W. Bush's tenure thoroughly discredited the Hamiltonian hard-power contingent. Obama seems headed toward a greater willingness to trust American power through negotiations, though he still shrouds exactly how he will use that power both to compromise and to make successful demands.

Promising for future policy debates, the realists of both the Truman-Acheson variety like senators Joseph Biden and Sam Nunn and the Republican realists such as Kissinger, Baker, and Scowcroft are receiving a fairer hearing once again. These realists have more in common with one another than with the liberals and conservatives of their own political parties. American foreign policy would profit if they backed one another more and their political parties less.

There's an opening now to strengthen the concept and the operations of American power. The first step is to get a realistic grasp on the new structure and distribution of international power. The second is to figure out how to work with this new power distribution and the limits it places on America's still formidable power.

CHAPTER 4

The New Pyramid of World Power

International power is not flat; it's pyramidal. The lion's share of the power sits in the top tiers, disproportionately with the United States and a second tier of major powers. But power is also dispersed below to unprecedented and complicating degrees. The result has been piles of dangerous stalemates, which can be fixed only by new approaches to organizing and managing power.

The history-enders are at it again. Twenty years ago, they said history had ended because America had triumphed over the Soviet Union, leaving America dominant and undisputed by any nation or ideology. History was finished because the United States could now simply dictate it. Not many years later, new post-history buffs are trying their hand for the opposite reason. They say history is over because the world has caught up with the United States, thus leveling and equalizing power. The United States is not nearly strong enough in the twenty-first century, they contend, to make history, shape great events, or solve major international problems.

But the history-enders are wrong: History continues because power and power struggles continue to rule the international arena, and because the fate of nations and peoples still hangs in the balance. History is still up to its usual tricks and mysteries, including an intriguing mix of old and new power configurations and instruments. This condition should be pushing us to figure out what's new and old in global politics and how to exercise power in this ever-unfolding world. But for a variety of reasons, that's not what we're doing.

There's never been anything quite like the world of this century, and the changes can be befuddling. That uncertainty is especially stark compared with the relative stability of the Cold War and the dependable balance of power that existed in Europe from the sixteenth through the twentieth century, when imperial maneuvers were disturbed by only the occasional Napoleon or Hitler. Les Aspin, Bill Clinton's embattled first defense secretary, expressed his exasperation about getting a handle on events as follows: "Even the experts couldn't make up what's going on out there."

Understanding what's happening in the world has been additionally obscured by several fashionable but misleading metaphors about the structure and distribution of international power. For one, many insist that the world is flat. But while it surely has flattened somewhat, it most certainly is nowhere near flat. Though power is now more dispersed than ever, the disparities in power remain vast and stark. For another, some argue that the world is nonpolar. It would be far more accurate, however, to see the world as a unique blend of unipolarity and multipolarity—with the United States clearly alone at the top, and with many other nations possessing highly varied degrees and kinds of power.

Today's world is neither flat nor nonpolar, but pyramidal: The United States stands alone at the pinnacle, with formidable and unique global powers of leadership, but not the power to dominate. Stacked below are many tiers of states, with most power still concentrated at the top tiers among relatively few states.

This distribution of power in the world is pretty clear. What's not clear is how to use power effectively in this new pyramidal, unipolar, multipolar context. So few nations have the power to get things done, and so many have the power to delay and resist. In such a universe, it becomes particularly difficult to solve major problems such as trade logjams, nuclear proliferation, terrorism, and global warming. Many sores linger and fester. Also, many of the most serious problems occur within nations, beyond the reach of even the strongest countries. Most problems now are less susceptible to good old-fashioned and decisive military force, and more amenable to the less visible and slower tugs of economic and diplomatic power. And many problems persist, interestingly, because the newly empowered countries of Asia are not yet ready to use their new powers as aggressively as the old Europe once employed its considerable powers. For Washington to wield power well in this universe requires a profound and creative understanding of these new dimensions of power, starting with the pyramid at its center.

IN THE NEW PYRAMID of power, the United States stands as the only country capable of global action and leadership on almost every major issue. There are very few situations, however, where Washington can prevail on its own. There are now precious few Panamas where a president can launch an attack, win decisively within weeks, and install a new and friendly government. As for diplomatic triumphs, getting Libya to abandon its weapons of mass destruction resulted as much from Colonel Muammar Qaddafi's own desire for more economic breathing room as from American pressure.

Nonetheless, there should be no doubting America's paramount position. Its economy outstrips all the other individual economies and is surpassed only by the entire European Union. China and India will take decades to catch up, if they ever catch up at all. While America now has competitors in technology and technological innovation, it remains the leader in those areas as well. And its military superiority far surpasses its economic advantages. The United States spent about as much on its armed forces in 2008 as all the other major industrial nations combined. More tellingly, it is in a class by itself in terms of usable military technology—the mix of hardware, software, and organization. On the diplomatic front, almost all countries turn to Washington, happily or otherwise, and regardless of whether they ultimately follow Washington. Only the

United States can act anywhere on virtually any military, economic, or diplomatic front. And in most parts of the world, the United States is also the sole guarantor of regional balances of power: for many Asians against China, for the Europeans against Russia, and for the Sunni Arabs against Shiite Iran. Despite these facts, America's preeminence is now regularly challenged by foreign policy experts and our own intelligence community. An intelligence study in late 2008 held that the United States has lost its position of dominance and is likely in fifteen years to become merely first among equal major powers. But this study doesn't really support the claim that the United States ever really was dominant and doesn't explain what the United States could do in the coming decade to avoid the loss of its central leadership power, which is at the heart of its power in the twenty-first century, as distinct from any prior era of mythical dominance.

The weight of the evidence is that America remains at the top of the international food chain and has unique powers to lead with regard to the most important and toughest international issues—ranging from nuclear proliferation and security to issues of trade, environmental issues, failed states, and health issues such as HIV/AIDS and pandemics. No other nation can play this role. No collection of other nations can play this role.

The second tier of countries consists of China, Japan, India, Russia, the United Kingdom, France, Germany, and just barely Brazil. Call them The Eight Principals, or simply The Eight. If Washington is the sole leader, they are the principals or managing directors of the global realm. Their views generally have to be taken into account globally and on many regional questions, but their economies and military capabilities do not permit them to take decisive or leading roles, either individually or collectively. They don't delude themselves by thinking they are equal to the United States. In many respects, they are more regional than global powers. But each possesses enough power to provide essential support to joint efforts with the United States and to block or seriously impede action by Washington.

All eight have narrowed the economic gaps between themselves and the United States and are now competitive with America on many economic fronts. Western Europe and Japan have counted economically for some time, but their economies are not as dynamic as America's. As for China, India, Brazil, and Russia, they have only very recently come to the fore in terms of trade and investment. Because of their dynamism and their relative novelty as economic players, experts usually overvalue their economies and their economic power. For example, China's GDP was roughly half of America's by 2008 estimates, and all the rest are smaller. For all of America's current economic ills, it remains the biggest and best market and a model of relative stability and continued potential for investment. It is Washington, not The Eight, that continues to grip the reins of leadership in organizations such as the World Trade Organization and the World Bank.

Militarily, The Eight are simply not first-tier powers. Neither alone nor jointly can they deploy or sustain significant military action beyond their borders. Only the United States can. All except Japan, Brazil, and Germany have nuclear weapons. China and Russia cannot be defeated in their homelands with conventional weapons, but by the same token, they can themselves apply decisive nonnuclear force only on or near their borders.

The diplomatic power of The Eight derives mainly from their economies and, in particular, their global trading and investment relationships. The size of their economies gives them a major role in worldwide negotiations on trade, energy, and the environment. Their financial and trading activities also make them vital to the success of economic sanctions. Simply put, no economic sanctions can be effective without them. If Burma can still turn to China for economic support, its dictators can survive a cutoff by almost all other nations. Washington can make economic life difficult for Iran as punishment for its pursuit of nuclear weapons, but as long as Tehran can trade with Russia, China, Germany, and Japan, it remains viable. No amount of American economic and diplomatic pressure on North Korea could have stayed Pyongyang's hand on its nuclear programs as long as China remained in Pyongyang's corner. Many of The Eight also benefit diplomatically by having vetoes as permanent members of the UN Security Council.

Of the lot, China has the greatest potential to become a global power. Even today, it has clout in many African and Latin American countries because of its power to buy local resources and make local investments. Overall, however, China and the rest of The Eight do not play an active role in diplomacy worldwide. They mostly counterpunch and complain—sometimes justifiably—about Washington, and about whoever sits in the White House. Mostly, they wait for Washington to organize action they think will fit their interests.

There is a third, narrow band of oil-producing states—the Oil and Gas Pumpers—that includes Saudi Arabia, Iran, the smaller Gulf states, Venezuela, and Nigeria (and obviously Russia as well). Their power derives from their large share of the global oil and gas supplies and the investment clout of their profits. They are essentially Enablers, helping to make things happen at home or abroad. The Saudis could use their money to buy weapons for the Afghan mujahideen to fight the Soviets. Iran can fund Hezbollah radicals in Lebanon. President Hugo Chávez can finance his populism at home and throw a few petrodollars to needy potential supporters elsewhere.

Iran remains a major producer, but it is a much larger and poorer country than most of its Arab neighbors, so its power is as an exporter and not as an investor. Russia also fits in this third tier (as well as in the second) now that it is the second-largest oil producer and largest natural gas producer in the world and turns profits into investments downstream in Europe. Nigeria has copious oil income and overwhelming problems at home.

For the most part, the oil-tier Enabler states stick to counting their profits rather than leveraging their resources for political bargaining. On occasion, Arab oil producers do make noises to encourage Western countries and Japan to be more sympathetic to Palestinian demands. Moscow has also begun flexing its energy muscles on diplomatic fronts. In sum, their oil resources and their liquid money have won them everyone's ear and a serious hearing of their concerns. That remained true, even in the financial crisis of 2008 attended by falling oil prices. The general assumption was and remains that those prices will rise again as national economies recover.

The fourth tier consists of mid-level states with mostly localized potential as Regional Players. This group includes Mexico, Nigeria, South Africa, Pakistan, South Korea, and Taiwan. Most are far behind most of the top three tiers economically. Some, such as Pakistan and South Korea, have substantial military strength for self-defense. If a problem is in their own backyards—Afghanistan for Pakistan or North Korea for South Korea—they will have an important voice in regional diplomatic parleys, but not usually a decisive one. Washington is still the leading and indispensable negotiator on North Korea and Afghanistan. Though Taiwan has a decent-size economy, it will continue to depend on America for its security—and that factor will ensure Washington's major role in Taiwanese-related issues. Pakistan, which also has nuclear weapons, exercises basic control of its own affairs, be they internal issues or fighting the Taliban within its borders, despite its economic dependence on America. Islamabad's power in these respects derives from Pakistan's size, location, and technological sophistication in certain military areas, but also, paradoxically, from its vulnerability to terrorists or extremists. This weakness permits them to deflect Washington's efforts to shove it hard toward internal reforms.

The fifth tier—which can be classified as Responsibles—encompasses as many as fifty states, medium and small, all over the map. Most are responsible world citizens such as Switzerland, Norway, Singapore, Botswana, and Chile. Many can mostly care for themselves and tend to their own needs, but don't cut a lot of ice with the major powers. They generally neither make nor submit to demands.

The sixth tier—the Bottom Dwellers or Problem States—includes about seventy-five states in varying degrees of political or economic disarray, or both. Their internal messes and conflicts sometimes impel top-tier countries to send in the troops (Afghanistan and Bosnia), sometimes give humanitarian relief (Bangladesh or Indonesia after natural disasters), and sometimes apply diplomatic pressure to combat human rights violations (Burma and Zimbabwe). They attract large-scale attention from major powers only when they fall apart internally or menace their neighbors. Examples include Sudan, Chad, the Democratic Republic of the Congo, Bosnia, Afghanistan, Uzbekistan, Nicaragua, and Burma. Many are the scenes of civil wars and ethnic cleansing; these states also harbor terrorists, engage in cross-border violence, and flirt with economic collapse, producing refugees and health issues that have the potential to afflict others. Such states have an international voice and even modest power when their internal woes become so threatening to others as to allow them to lay claims to international resources. Some, like Sierra Leone, which recently concluded a civil war, have required and received outside peacekeeping and economic assistance to prevent a resumption of fighting. Some, like Darfur in Sudan, where the situation is awful, touch Western humanitarian hearts or trigger fears of terrorism, but don't levitate to become more than sad political topics in top-tier countries.

The nations in the last several tiers also extract bits of power from the now widespread practice of multilateral diplomacy where the practice of consensus reigns. In forums like world trade, global warming, and health, they have a voice. That's because the expectation in these multilateral arenas is that every nation should be a player and a signatory. Nongovernmental organizations (NGOs) often help add weight to the views of these bottom-tier countries.

The seventh and final tier consists of the Non-State Actors. They include refugee and human rights advocacy groups (the NGOs), terrorists, the international media, and international business. They are a highly disparate bunch in interests and actions, and they often act in ways contradictory to one another's interests. They are now thoroughly intertwined with governments, societies, and individuals all over the globe and operate worldwide. It's difficult to measure their influence, but they dwell everywhere and usually manage to get at least a hearing on big issues and a real voice where their expertise is engaged.

Many NGOs have long operated within countries, but never approaching their present numbers or influence. Global communications and the Internet magnify their views as well. Perhaps no twenty-first-century business touches the power of the East India Company, but there are now tens of thousands of such companies, small and large. These firms have considerable impact on issues directly affecting them. It would be obscene to list terrorist and extremist groups among the NGOs, but they are, nonetheless, Non-State Actors of critical importance.

To summarize, the distribution of power in the pyramid looks something like this: the United States uniquely has the power to lead, but certainly not to dictate; the second tier, The Eight, can be either the principal partners or definitive blockers of Washington. All other states in the other layers have sharply varying powers to resist or to help or hurt in subordinate roles. And one final point about this pyramid: For all the enormous disparities in influence and the continuing power advantages of the top tiers, the entire system tends toward stasis, inaction, and drawn-out pulling and tugging.

IN OTHER WORDS, THERE are some good reasons behind the erroneous conclusions of the world-is-flat crowd. They rightly highlight a slew of historic shifts that *have* reshaped the distribution and composition of international power—the geographic transfer of power from Europe to Asia, the decline of military power and concomitant rise of economic power, the attendant splintering of power accompanied by a growing interdependence between and among states, and the sprouting of threats from within rather than between states. Their mistake is in taking their conclusion about the leveling process to extreme lengths. This process has empowered many states that were previously insignificant. But by no measure has it eliminated the substantial disparities in power—let alone the centrality—of power itself, and the power of the United States.

The geographic shift of power from Europe to Asia is the most noted of these phenomena. Europe ruled the roost for almost 500 years, militarily and economically. Now, Asia has most of the world's dynamic economies—particularly China, India, and parts of Southeast Asia. And the military capabilities of China and Japan rival or exceed those of the major European countries. Of course, Russia, the United Kingdom, and France have nuclear weapons, but so do China, India, and Pakistan.

Paradoxically, this power shift has tended to magnify America's global power rather than diminish it. That's because as of this writing, Asians have played a less assertive diplomatic role globally than did Europe. The major states of Europe historically exercised their power globally. Even today, with their power greatly dimmed, they still make that effort and merit a seat at most bargaining tables. For the most part, however, Asians have thrown their weight around in Asia, and on economic issues. Generally, China and Japan, the two Asian powers most capable of distant intervention, refrain from involvement in interstate conflicts on other continents and usually take an arm's-length position on the internal politics of faraway states. India has been even more reluctant than Beijing to seek influence in distant conflicts. Moreover, Asian-Pacific groupings are still in their infancy and have less clout than the European Union and its affiliates.

Europe, of course, could weigh in more heavily on future power scales if the ever-expanding European Union were to develop a single defense, economic, and foreign policy. But that goal remains distant. The British and French, in particular, will still insist on their policy independence. In the meantime, European diplomacy leans toward the fluffy and reactive.

In perhaps a decade or more, major Asian states may feel safer on their home fronts and permit themselves the luxury of greater involvement beyond their continent. For now, however, a weaker Europe and a circumspect Asia allow more running room for Washington.

AS THE GEOGRAPHY OF power has shifted, so has its composition. International power now has more of an economic flavor than ever before and an enhanced diplomatic dimension, as well as a reduced military component. Internal economic strength, to be sure, has always mattered and has formed the basis of almost all external power. But military force had been the principal calling card, while diplomacy has been trained on convening alliances or counter-alliances, and putting the final touches on victories or defeats. Major powers went to war to acquire territory and riches. By contrast, governments mainly limited their economic actions to protecting trade and collecting tributes from their defeated adversaries. Private banks and traders went their own private ways, calling in favors as needed from government. For the most part, empires and nations were not nearly as involved in one another's economies and daily economic lives as they are today. The range of interstate economic transactions has expanded enormously.

Now, national power finds its most common expressions in a state's capacities as a buyer, a seller, a lender, a borrower, an investor, an innovator, and a benefactor. In most countries today the government's role in these activities is substantial. On the whole, however, economic activities are pursued for economic ends, and economic power is used to exact economic concessions. From time to time, of course, and depending on the issue, nations do link economics to strategic and foreign policy aims. The major focus of diplomacy in the twenty-first century will be on economic transactions, while military force will usually wait in the wings.

Most governments allow for a great deal of economic give-and-take with other nations and foreign businesses. They even show some tolerance for setbacks at economic bargaining tables—especially compared with the zero tolerance they have for security disputes. Nations abide one another's economic activities now without resorting to war or military threats. Nations seem to live relatively easily, if not happily, with the interdependencies and the attendant economic ups and downs.

Washington still sits at the center of the world economy, in the varied roles of both leader and mere bargainer, supplier and consumer, the main engine of worldwide economic growth, and the closest thing there is to a manager of the global economy. It thus has the primary responsibility for managing the global trading and investment systems, and it holds most of the keys to the international economy, through its membership in such entities as the World Bank, the International Monetary Fund, and the World Trade Organization. It is hard to imagine other nations digging their way out of their economic doldrums or even thriving without U.S. support, approval, or leadership.

Oracles of the theory of the decline in U.S. power have disputed these propositions for some time, and did so far more vigorously after America's economic crisis blossomed in 2008. They saw the crisis as proof of the deepening rot in the American economy and argued that the new economies of the world, such as those of China, India, and Brazil, would be insulated from the American financial disasters. They advanced the decoupling argument, namely that these new economies had accumulated such large currency reserves from export profits and had built such strong banking systems as well that they would survive America's credit fiascoes and even thrive relatively. They maintained that the crisis demonstrated that a major shift in world economic power had occurred. But these declinist assertions have been proven to be weak and shaky already:

First, though the American economy was perhaps hardest hit by the 2008 economic earthquakes, the others have not been insulated and have suffered almost as badly. Second, investors from around the world took refuge not in these new currencies, but once again in the dollar. Again, no other economy and no other currency were considered as safe or safer than America's. Third, non-American leaders, new and old, of virtually all nations did not grab the leadership reins to organize global rescue, but as always, they turned to Washington and only to Washington. It was a Bush administration already on its knees politically that stepped out front to coordinate international interest rate cuts. And it was lame duck Bush, not Chinese or Indian or European leaders, who summoned presidents and prime ministers from the world's top twenty economies for an emergency conference in Washington in November 2008. Fourth, and perhaps most tellingly, the leaders of the newly empowered econo-mies have not been running around demanding that Washington be replaced as the economic leader, let alone claiming this position for themselves. They don't want that responsibility, and they don't have the power or position to carry it out. They know better than America's fashion designers that they cannot fulfill this leadership role. What they want and what they will have is a presence and a voice at every international economic table, commensurate with their new standing.

The financial crisis shook the entire pyramid of power, including America's lonely perch at the very top, and rejiggered the economic balance of power. America counts somewhat less. So do Japan and Western Europe. Several of the new economic giants are much stron-ger both relatively and absolutely, and count for much more than they used to. But together and separately, none can match America's economic power and position overall, nor replace Washington as the global economic leader. The pyramid still holds, despite the continuing crisis, and America—for all its new woes—remains the world's only economic leader for the foreseeable future.

As military power declines in importance along with its short-term and often decisive effects, and as economic power rises with its longer-term and more complicated results, the overall diplomatic terrain has become highly cluttered, complex, and often stalemated. There seems to be an unprecedented amount of diplomacy on an ever-burgeoning set of agendas—but with very thin results. It seems that the will to make concessions in order to reach agreements has waned. Rather, governments are now inclined to live with disputes and differences and operate on an ad hoc basis. Political leaders are more reluctant to concede unless the compromise is essential to their political position at home. The fair conclusion seems to be that the new distribution of power and its altered composition have slowed down and even knotted up the process of settling differences.

The rise of multilateral diplomacy on top of the traditional bi-lateral variety has raised additional barriers to successful diplomacy. Between 100 and almost 200 nations now need to be shepherded toward consensus on all the hot topics, such as trade and health. Nothing in history approaches the current magnitude and complexity of multilateral diplomacy. And like all the other changes in the disposition and composition of power, it slows down the resolution of conflicts.

THE POWER OF THE strong has also been choked by an unprecedented degree of interdependence among states. Economic interdependence is palpable. One telling example is the highly entangled Chinese-American economic relationship: for its economic growth, China depends on exports to America, and America depends on China's use of its profits to purchase U.S. securities to finance those imports. Or take Western dependence on Middle Eastern oil, and the Arab oil producers' need for Western military protection and relatively safe investment opportunities. But the condition of being thoroughly intertwined also shows up in national security as well. Specifically, some of America's most advanced missiles and communications systems have parts manufactured in a dozen different countries, including countries with less than entirely friendly relations with Washington.

This interdependence means that deals take longer and are more difficult to finalize than in decades past. Issues drag on. And contrary to the bright expectations and violin music that usually accompany toasts to globalization, the phenomenon has tended to freeze power relationships and perpetuate the status quo. By contrast, the two great benefits of interdependence are that many countries are becoming richer, and nations aren't killing each other over their disputes.

Looked at solely in power terms, mutual dependence has reinforced the splintering of power. Even small and midsize states have little pieces of the action, if only the power to say no regarding their own affairs. It is this splintering of power that makes the world look flat to the naked eye.

Finally, international power is stretched and thinned by the new source of the most serious threats to international peace and security. Such threats now spring more from within nations than between them. This makes the threats harder to reach and more difficult to combat. It was easy to chase Saddam's army out of Kuwait. Later, it was relatively easy to defeat the Afghan Taliban on the battlefield and occupy Kabul and beat Saddam's army and occupy Baghdad. The quagmires came later, from insurgencies within Afghanistan and Iraq.

The most worrisome threats now stem far more from terrorists acquiring nuclear and other weapons of mass destruction than from traditional conventional armies. Nations like Iran can lose wars and later strike back with terrorists against the homelands of the victorious states. The United States and Israel can theoretically destroy Iran's nuclear capability and find themselves faced with terrorist attacks on their own soil. Or weak states fall apart and visit their problems on neighbors near and far. Or weaker states such as North Korea can acquire nuclear weapons and change the balance of power with this single act. This is why the acquisition of nuclear weapons, or even the relevant technology, by—say—Iran, is regarded as the contemporary equivalent of an act of aggression. Or a state can collapse and drown its neighbors with refugees and other woes, as often occurs in Africa.

What we've seen is that if leaders of these weak states think that outsiders are trying to dictate or meddle in their internal issues, they dig their heels in even more. Not surprisingly, they define internal issues very broadly—running from how they rule their citizens to their decision to pursue nuclear weapons. It's easier for Washington to get foreign leaders to stop bothering their neighbors than to get them to lighten up on their own people. China will work with Washington on North Korean nuclear weapons, but turns a deaf ear to pleas for human rights and democracy within its borders.

The pyramid of power encompasses all these changes, but the overall effects are far from linear. A final gaze at the pyramid reveals two competing crosscurrents: One is the greater splintering of power at mid- and lower tiers, and the other is its increased concentration in the upper tiers. How these crosscurrents are managed by policymakers, particularly those in the United States and The Eight, will be one of the most critical factors in determining where history takes us next.

The first current, the splintering of power, began in earnest after World War II with the multiplication of nation-states and with their new nationalist fervor to resist outside pressures. Experts argue over whether this phenomenon has been good or bad, and the answer seems to depend on where one sits. To those newly empowered, it seems a boon. The major powers can't push them around nearly so readily as before. But to those interested in some semblance of world order, the fractionalization of power is producing a world sinking in deadlocks. It is becoming increasingly difficult to solve problems and resolve conflicts.

This splintering, a phenomenon unprecedented in history and laden with unforeseen consequences, mesmerized policymakers and policy experts. They became so absorbed in what was happening at the middle and lower ends of the power pyramid that they scanted something at least equally potent: the current evolving at the top.

More and more, power was being concentrated in the top three tiers of the pyramid. The paramount power of the United States at the very apex should have been perfectly clear to all. The economic, military, and diplomatic facts demonstrated beyond argument that the United States was the world's sole leader on major problems, yet not a dominant power. But these facts were obscured by what most perceived as the unilateralist rhetoric and actions of President George W. Bush. This unilateralism tended to isolate America, thereby diminishing its power. Under Bush, there was often no lead for nations to follow. President Clinton had not fared a great deal better on the global leadership front, although he was far less unilateralist and isolated than Bush. In any event, neither understood the nature of America's superior power and unique leadership role, and therefore both failed to take advantage of it.

Nor did The Eight in the powerful second tier see the new role they could play, given their new interests and power. They saw they had new power, especially in China, Russia, India, and Brazil. But leaders in Beijing, New Delhi, Moscow, and Brasília did not delude themselves about how their power compared with Washington's. They grasped that their power rivaled America's but did not equal it. Unlike America's flat-worlders, they did not exaggerate their power and overreach. In fact, most underreached and did not stand up for what they themselves saw as their new interests in sustaining world order. They had a much larger stake in world economic affairs, and understood that this required a decent amount of world stability. But of The Eight—China, India, Brazil, Japan, France, Germany, Russia, and the United Kingdom—few have been willing to put their power on the line to help others solve problems or to resolve conflicts. The third-tier countries, mainly the oil-rich ones like Saudi Arabia and the Gulf states, also have not distinguished themselves in these respects. And so we all drifted on and about these last two decades.

Policymakers and analysts have been wallowing in the new and critically important splintering current and not zeroing in nearly enough on the power-concentrating current. The consequence of this misjudgment has been most serious: It has been to reinforce the stalemating effects of the decentralizing current. Since policy leaders seemed unclear and uncertain about how to use their powers to solve the problems, they essentially went along with or gave in to them—and the problems have deepened accordingly. In the last analysis, we have all failed to think creatively about how to productively combine power at the top of the pyramid to get things done.

The power pyramid reveals the stalemate problem we've been living with as well as the policy solution. The solution is to forge cooperation within the top two tiers in particular. The record is quite plain on this.

When Washington and The Eight want to right a dangerous wrong, they can. Saddam's invasion of Kuwait was a threat to world order, and they joined together successfully to chase his troops out.

When they want to tamp down a dangerous common threat, they can. Look at the progress they've made together through cooperation on intelligence and law enforcement operations against terrorists. If they want to combat the threat of nuclear proliferation, they have a good shot. With a united diplomatic front, most of The Eight moved North Korea, one of the most intransigent countries in the world, along a policy path to curb its nuclear program—at least reaching a paper agreement, if only fleeting. On Iran, where they have not been nearly united enough, the dangerous nuclear issue never got nearly as far.

Obviously, there are roadblocks to such cooperation, not least of which are the domestic politics of these major powers and how each perceives its own national interests. And obviously, cooperation sounds nice and is nice, but is often hard and costly to arrange. But the overriding reason for cooperation is clear: It's the only way to solve common problems. And the top-tier nations have the power to succeed if they rethink how to manage their power.

The critical step is for Washington to see this light—to master the power pyramid and rethink how to use power in this new unipolar and multipolar context. It means fashioning a coherent overall strategy, an essential and torturous step our leaders rarely take. It means developing genuine understanding and knowledge of the internal politics of other countries; knowing what we can reasonably ask of our intelligence experts; doing a better job of managing the politics of foreign policy in America; and getting a much better handle on the new forms of military, economic, and foreign policy power. It calls on us to relearn the connection between power and policy. The answers reside in the pyramid, along with the new rules for exercising power.

Part II

Rules for Exercising Power

Part II
Rules for Exercising Power

CHAPTER 5

STRATEGY AND POWER:
MUTUAL INDISPENSABILITY

Good strategy—setting achievable goals and priorities, knowing your power sources, and sequencing moves—is the essential starting point for exercising power. And when you do this, you will see the futility of unilateral action and the necessity for power coalitions to solve problems.

Multitudes of Americans were baffled when American leaders proclaimed the mighty powers of Iraq's Saddam Hussein and Iran's President Mahmoud Ahmadinejad. They weren't mighty. Calling them mighty exaggerated their power, while diminishing American power. An unhappy but instructive precedent can be found in the year 1960. These are all tales in how not to make strategy by giving away power.

"Missile Gap" was an oft-reprinted banner headline that year, as the Democratic presidential candidate, John F. Kennedy, criticized the incumbent Eisenhower administration for allowing the Soviet Union to gain superiority over the United States in the number of nuclear-tipped long-range ballistic missiles. This charge contributed to Kennedy's narrow victory over Richard Nixon, Eisenhower's vice president, in the election. And indeed, there *was* a missile gap: the Soviet Union had by some accounts fewer than ten, while the number for the United States was in the thousands, to say nothing of America's equally vast advantage in nuclear-capable long-range bombers. It remains uncertain to this day whether Kennedy's grossly inaccurate overstatement sprang from sheer ignorance or cynical political calculation. Whichever it was, he corrected the record only after winning the election—after the damage was already done.

In fact, Kennedy played right into what I call Moscow's Wizard of Oz strategy. He gave the Soviets free power, power they neither had nor deserved, as they hid behind their iron curtain, concealing their weaknesses and continuing to bluster as if they were America's equal or superior.

The proof that the Soviets were, indeed, using a Wizard of Oz strategy was that they never cried foul about the "missile gap." They never once said, "Hey, wait a minute. We're holding the short stick, just struggling to catch up with you." To the contrary, Communist Party head Nikita Khrushchev embraced the myth of the "missile gap," acted as if the Soviets actually had superiority, and publicly warned the West, "We will bury you." The Soviets' game plan was to mask their inferiority, husband their limited resources, divert U.S. energies to defending vulnerable friends and allies, and steer clear of losing confrontations—all to stay in power at home, attain even greater power in Eastern Europe, and play the role of superpower.

The free power from Washington was a gift. The Soviets knew they had the weaker hand, and so, when Americans declared Soviet military superiority and presented them with power on a platter, the Soviets were only too happy to help themselves. And the platters kept coming—just as they did for Saddam and Ahmadinejad, who also took the power and fed off it.

To induce anxiety about the Soviet Union among their fellow citizens, American hard-liners conjured up one gap after another: in missile defense, in intercontinental ballistic missile (ICBM) first-strike capability, in fighter aircraft, even, most implausibly, in medical care. Most hard-liners knew, I believe, that the Soviet Union was not militarily superior. American superiority was so substantial that they had to be aware of it. But they fervently believed that Americans wouldn't pay for additional defense expenditures without being scared silly. They were nearly maniacal about keeping Americans on their toes in the fight against communism and didn't seem to care that they were giving the Soviets free power. Today's hard-liners similarly believed their exaggerations of Saddam's and Ahmadinejad's powers necessary to justify a military assault against them.

I was embroiled in battles over the balance of Soviet and American military might for most of my career in and out of government. This issue, more than any other, branded one publicly in Washington as either a bleeding heart or a hardheaded realist. It aroused hard-liners because they correctly concluded that they had to win this fight in order to win other ones. If it were established that Moscow had military superiority, then the United States would have to spend more on defense and be tougher abroad to discourage Soviet adventurism against a weakened America. When I prodded James Schlesinger on this subject during his tenure as President Ford's defense secretary, he responded: "Well, I suppose the Soviets don't have military superiority in fact, but they have it in perceptions." To which I responded, "It is we who are creating those perceptions." Hard-liners are doing much the same today with Iran. Their exaggeration of the Iranian threat is laying the groundwork for exaggerated responses, which hard-liners believe to be necessary, and attainable only by exaggerating the threats.

Moscow bled American resources and power all over the world for more than forty years and stayed in the game longer than seemed possible on the basis of its own actual resources. The United States prevailed for many reasons, not least our free society, our vastly superior economy, and our strategy of containment. American resources are being similarly bled today in Afghanistan and Iraq, and Washington could be headed in the same direction regarding Iran.

There is nothing more central to the exercise of power than a good strategy, and the United States does not now have one. Without this, there can be no sense of attainable objectives and no plan to wield power effectively and blunt the power of opponents. Current offerings include these: We hear of an alliance of democracies (although there's little basis for their joint action), of Washington sometimes leading and other times letting others lead (who else can lead and on what issues?), of colluding quietly with Russia and China, of not working with these would-be adversaries under any conditions, of multilateral action (an oxymoron), and of leaving it to a stronger United Nations (an amusing notion, even to UN officials). Indeed, most of these proposed strategies are so flaccid as to suggest that the meaning of strategy itself has been put out to pasture.

THE GREEK WORD FOR strategy, *strategos*, defined as "from the office or command of a general," was too narrow a construction for what Niccolò Machiavelli contemplated in *The Prince*. He did not use the word strategy but instead talked throughout his manuscript about how to achieve success, and this, in turn, almost always rested upon clarity of aims, willpower, and doing whatever was necessary to achieve goals and extract obedience. He then went on to offer many rules for doing this, primarily being good at the art of war and keeping your political enemies off balance.

The Napoleonic French eventually gave strategy its current meaning as the art and skill of maximizing one's means to achieve one's desired goals. Henry Kissinger, a twentieth-century master of power, has exalted the importance of strategy above all else. "[H]istory demonstrates," he has written, "that superiority in strategic doctrine has been the source of victory at least as often as superiority in resources. . . . Thus the key to a proper doctrine is the correct understanding of the elements of one's superiority, and the ability to apply them more rapidly than the opponent."

Kissinger particularly admired Otto von Bismarck, architect of the German Empire and (from 1871 to 1890) its first chancellor. His goal was to unite the Germans, despite the opposition of Europe's major powers and with the likelihood of having to go to war against them. But he knew well that little Prussia could not reveal its ultimate goal or confront all its enemies at once.

His dream was a united Germany on a par with Russia, Austro-

Hungary, France, and the United Kingdom. But the emperor in Vienna—who held sway over a number of unruly Germanic multitudes—along with other European emperors, didn't want a united Germany. These rulers preferred nonthreatening disorganization in the geographical center of Europe. What's more, the conservative rulers in Moscow and Vienna trembled at anyone's igniting nationalism in their midst, lest it become a fire that would engulf their own cardboard dynasties. To unify the various Germanic fiefdoms under the king of Prussia, Bismarck needed a scheme to overcome the power and suspicions of these established empires, and a way to leverage Prussia's assets, namely, its political cohesiveness and its fine army.

The Prussian diplomatic genius soon found the necessary leverage, by exploiting the inability of these cumbersome empires to mobilize against anything short of an imminent, direct, and dire threat. So Bismarck calibrated the use of Prussian power accordingly: he masked his desire for the German Empire and kept his threats well below imperial thresholds. He spun his web by serially isolating each opponent diplomatically, provoking each into appearing to be the aggressor, then unleashing his extraordinary Prussian army, which crushed its enemies with brief, sharp blows. This allowed Prussia to absorb one limited piece of strategic territory after another without threatening the survival of any of the emperors. First, Bismarck gained Vienna's assent to wage war on Denmark in 1864 and thus to acquire the German duchies of Schleswig and Holstein. Then, in 1866, he calmed Russia and France as he agitated Vienna into a showdown on the battlefield, knocking Emperor Franz Josef into submission and dissolving his hold on southern Germany in Prussia's favor. Finally, in 1870, with Moscow and Vienna as distracted onlookers and with London absorbed with its colonies, Prussia struck and swiftly humbled the French armies, annexing the predominantly French Alsace-Lorraine against Bismarck's wishes (he feared creating French revanchism). Shortly thereafter, the king of Prussia was crowned emperor of Germany as well.

Bismarck had created an economic and military powerhouse in

the center of Europe, but he did not then try to muscle his neighbors with his newfound superiority. He rightly feared that if Germany overstepped its bounds, the others would unite and defeat it. His successors thought they could dominate their European neighbors and get away with it. Instead, they destroyed the German Empire and Europe itself.

STRATEGY OFTEN LOOKS EASY to construct, but nothing is more difficult, especially in the twenty-first century, with its bewildering array of power centers and intractable global challenges. Americans are frustrated and uncomprehending when they stand by and see their nation's extensive power being thwarted again and again. This happens because no one has set forth a strategic kernel that harmonizes the complex realities of the twenty-first century with the needs of American national interests into a single, simple insight, one that both reflects the new global realities and provides an organizing principle for action.

The elements of strategy are widely familiar, yet they are seldom attained. The discipline required to examine all the hypotheticals and to weigh all the options always seems to take too much time away from meeting the daily quota of mistakes and to be too much trouble until it's too late.

The strategic enterprise begins with laying out goals, imagining nightmare scenarios, and assessing all threats, problems, and opportunities. All this must be boiled down to achievable objectives, priorities, and trade-offs among goals. There must be a rigorous accounting of the powers available to achieve the ends, which evolves in good measure from a brutally realistic appraisal of the strengths and weaknesses of all parties—above all, one's own. These are hard to frame, and even harder to mesh together.

More difficult still are two additional steps, which are much less obvious than the others and more challenging to construct. One is the "first door" process: the task of deciding what to do first to make all subsequent actions easier. The second is to know the "power source," meaning which factors will provide power in both general and particular situations. From all this can be derived the strategic kernel, or the essential element of the strategy that will inspire and drive the entire enterprise.

The initial step, obviously, is to define the attainable objectives. It sounds easy, but it isn't. The tendency is invariably to state the loftiest of goals, the most desirable and the grandest. Who isn't for world peace and the cessation of conflicts everywhere? Why expose oneself to domestic critics eager to pounce on the weak-kneed who are ready to compromise before the bargaining even starts? Alas, grand and unrealistic goals cannot be readily tossed into the garbage at night with only the cats taking notice. They accumulate constituencies and become entrenched. But they have to be fought off because they invariably produce a stalemate at best and a costly defeat at worst.

Achievable goals, by contrast, provide room to maneuver at home and allow the nation's power to be focused more pragmatically abroad. Washington lacked the power to transform Iraq into a democratic free-market paradise, and wasted enormous power on this pipe dream. But it is conceivable that had Washington simply focused its considerable energies on providing decent security in Iraq and on stabilizing the country, it could have reached this more modest goal already. Washington could try to simply stop Moscow's exercise of its new power and fail, or it could attempt to build a new strategic relationship with Moscow (as Obama seems to have attempted by declaring a "reset" of relations) and restrain it that way—and have some chance of success.

In sum, achievable goals make for achievements, and achievements make for power and the potential for attaining even more power. A fool is someone who defines a problem in such a way that it can't be solved. If it can't be solved, it shouldn't be pursued. If failure to pursue it exposes major weaknesses, then you must look for successes elsewhere, where you can achieve them and use them to compensate for your failures. The major example of this is, of course, the Nixon-Kissinger triangular diplomacy to cloud defeat in Vietnam.

Second, strategy requires setting priorities and making trade-offs among objectives. But the American way is to want it all. Washington wants Europe to put pressure on the Palestinians for a compromise with Israel, but refuses to give the Europeans a responsible role in the negotiations. It can't have both. Washington wants to deploy a missile defense system in Poland over the Russians' objections, and also to enlist Moscow's help on Iran. It probably can't do both. Putting objectives in some achievable order is the only way to establish cohesion between policy and power. Without it, power works at cross-purposes, and fails.

Setting realistic, achievable priorities also permits leaders to distinguish the absolutely vital from the merely important and the important from the marginal. It's obvious that all power and resources are limited, and that choices have to be made among competing goals. And yet, astonishingly, politicians and pundits alike declare virtually every world event to be a historic crisis, and drag presidents into needless messes. This happened repeatedly in the Third World during the Cold War, when Washington's operating principle should have been that no place where we couldn't drink the water could be deemed a vital interest. By contrast, the Soviets were brilliantly prudent in spending their limited power. While American leaders routinely proclaimed nine-tenths of the globe to be so "vital" as to require U.S. military protection, Moscow defined only one region as strategic (their word for vital), namely Eastern Europe.

Foreign policy priorities, when openly explained and sold publicly, give the president a decent chance to conserve time and resources, two of the most precious commodities in applying power. Presidents have to save their power for the big challenges. Compare Bush's decision to invade Iraq before he controlled Afghanistan with Truman's strategy of just pouring resources into Western Europe and Japan. The current Chinese leaders understand the importance of priorities and don't allow anything to divert them from promoting their internal economic growth and maintaining domestic stability.

Third, leaders must carefully assess their own nation's strengths and weaknesses, as well as those of their allies and adversaries.

Where are you strong and weak, and where are others strong and weak? American openness is both a strength and a vulnerability. How can the downside of transparency be minimized? Is the main weakness of America's adversaries such as China and Iran their economies or their lack of internal political legitimacy? How are these weaknesses exploitable without causing larger disputes? What constitutes a credible American threat? Did George W. Bush, for example, really believe that his threatening of the North Koreans would stop their nuclear weapons development program, and if not, why did he do it? If the successful application of U.S. power will take time, as it normally does, how can a president buy that time in a country with the patience of a two-year-old?

The fourth step in creating a strategy requires even more creativity than the first three: It is to choose the proper sequence of moves toward one's objectives, and, in particular, to decide which of the many doors to solving a problem should be opened first. Many say that on tough problems you've got to open as many doors as possible virtually simultaneously. But in practice, that can rarely be done.

Usually, pushing the most important door first will make it easier to open the others. Take China: Washington has no particular power to pressure Beijing to curtail its current military buildup or to cease its threats against Taiwan's independence. But Washington can enhance its mutual economic ties, and in this way, strengthen its main lever over Beijing. That, in turn, will allow the White House to better manage the serious security disputes. Or take Darfur. There have been many efforts to convince all parties—the Sudanese government, its allies, and the rebels—to make promises of good behavior. But the atrocities against non-Arab African ethnic groups in Darfur are driven by the Muslim Arab government in Khartoum, and nothing will be accomplished until that government is convinced it will pay a heavy price for continuing on its murderous course. Deciding that Khartoum is the most critical door and focusing all power there may be the only hope for a solution.

The "first door" procedure sometimes suggests taking on easily winnable fights before moving on to the more difficult ones. The idea is to build up success and power. For example, the initial step on Darfur probably should be gaining European and Chinese support for pressure against Khartoum. And there are times when the toughest fights must be joined at the outset. For example, Washington certainly had to make changes in its political strategy in Iraq—say, by developing ties with Sunni insurgents—before military security could take hold. Similarly, Richard Nixon and Henry Kissinger wanted the Egyptian president, Anwar Sadat, to make peace with Israel, and they understood that the first door to a stable peace would have to be Israel's agreement to return the Sinai to Egyptian control, and, barring that, to restore Egyptian honor and dignity after another military defeat.

The fifth step in framing strategy is to understand with great clarity the true source of one's power overall, as well as in each individual circumstance. Presidents have to clearly grasp exactly where American power comes from. President Nixon understood that establishing U.S. power in the Middle East required demonstrating that only America could shape a peace agreement between Israel and Egypt. Today, American power derives in good measure from the United States' being the ultimate regional balancer throughout the world—in Asia against China, in the Middle East against Iran, and in Europe against Russia.

Good strategy can produce extraordinary results. And yet, most experienced hands in government can attest to this: Leaders hardly ever do it. When disaster strikes, as it invariably does, officials are always left asking, "Why didn't we think through our strategy before we started down this path?"

I remember one of my last National Security Council meetings during the Carter administration. Its purpose was to prepare the president for his forthcoming visit to South Korea. We did the usual run-through of half a dozen issues such as nuclear weapons and human rights, looking at the options for each, with each member of the senior staff expressing his preferences. It lasted the usual two-and-a-half hours. When it was over, the dozen or so people in the room moved toward the door, only to find it blocked by Defense

Secretary Harold Brown. He stood with his hand on the doorknob, gazed at the ceiling, and said, "You know, we didn't discuss our policy toward Korea." The most basic issue had never been raised, much less carefully discussed. Everyone filed out.

When I was assistant secretary of state for politico-military affairs I often urged my boss, Secretary of State Cyrus Vance, to give strategy or policy speeches. It was still early in the Carter administration, but we were already besieged by critics who said that we didn't know what we were doing. I thought Vance could persuasively demonstrate to our critics at home and abroad that the administration did, in fact, have a foreign policy. Finally, frustrated by my repeated recommendations, the mild-mannered Vance snapped at me: "Policy is baloney." Much as I respected the man, and continue to respect his memory, policy is far from baloney. It is essential to a successful strategy for the conduct of American foreign affairs.

AMERICAN LEADERS HAVE, ON occasion, understood how to develop strategies, and on three critical occasions, they designed brilliant ones that made major contributions to the winning of the Cold War.

President Harry Truman focused on a strategy of helping key allies more than responding to Soviet military strength. Then there was the Nixon-Kissinger strategy of snatching victory from the jaws of defeat by demonstrating America's unique diplomatic powers. And finally, there was George H. W. Bush, James A. Baker, and Brent Scowcroft's strategy of strengthening an adversary to facilitate his capitulation.

The Truman team's strategy marked the golden age of U.S. foreign policy, as glorious in our history as the founding fathers' creation of the Constitution. The key decision by Truman's team was not to negotiate with Moscow from weakness but instead to focus on building the foundations of strength. With America's demobilization after World War II, the Soviets enjoyed massive superiority in conventional forces. This put them in no mood to make compromises. The second step was to see that America's biggest postwar vulnerabilities were also its greatest sources of potential strength, especially the potential of Germany and Japan to bolster American security. So, Truman used America's clear area of superiority, its economic strength, to help restore these shattered countries economically and to stabilize them politically. This process of building up Germany and Japan was Truman's "first door," and it was achievable. The United States had the economic wherewithal to do it, and Truman persuaded Congress to pay the bill for the Marshall Plan. The United States did not require Soviet cooperation to do this, nor did it have to compromise with Moscow to accomplish it. It could take these actions on its own. The final step was to complete and create new international institutions through which American power would be exercised. These included NATO, the United Nations, the World Bank, and the General Agreement on Tariffs and Trade. Through these new, postwar organizations, America would lead and set the directions, but others would be able to have their say. Truman and his national security team did not believe that multilateral action would lead to weak actions; on the contrary, they were convinced that this was the means to exercise power most effectively—and they were right. In sum, the Truman team understood that for the foreseeable future American power would be based on its ability to strengthen and protect friends, rather than to defeat enemies.

Nixon and Kissinger had to struggle with a hitherto unfamiliar problem for Americans: the prospect of defeat. They tried to stave off defeat in Vietnam, but had probably foreseen its inevitability. Nonetheless, they upped the ante by announcing repeatedly that an American defeat would turn the United States into a "pitiful, helpless giant." This frightening image was designed to taint their domestic adversaries as defeatists and to silence congressional demands for faster U.S. troop withdrawals.

Most American policy experts believed in the domino theory, which held that failure to stop aggression in one place, invariably some small place, would lead to more expensive and dangerous challenges in more consequential places. Experience in World War II buttressed the domino theory: Hitler and Hirohito had picked off smaller countries without international penalties, and this experience served to convince them that the Western powers didn't have the stomach for a fight and emboldened them to reach for more. The psychology of power also sided with the domino theorists; appeasement and docility most assuredly do encourage bullies and discourage friends.

Nixon and Kissinger were not going to wait for the dominoes to fall. Instead, they conceived and carried out a brilliant three-ring diplomatic circus—one that blunted and diverted the impact of impending defeat in Vietnam while demonstrating America's peerless diplomatic power.

The first ring was triangular diplomacy with the United States as the pivot between the Soviet Union and China. Only the United States could play pivot. Beijing and Moscow were more at odds with each other than either was with Washington, and neither would have permitted the other to be the pivot. Nixon did not win significant concessions from either of them on any issue, and especially not on Vietnam. But he did demonstrate that Washington was the first among the three world giants. As Kissinger wrote in his memoirs, Washington's primary task was to produce "conspicuous successes" for the American people and the world audience at this moment of national distress—and it did exactly that.

Second, Nixon and Kissinger put American diplomatic power on display in the Middle East. They moved with dispatch to capitalize on the ongoing 1973 Yom Kippur War, after Egypt had attacked Israel and as Israeli forces were gaining the upper hand. Nixon and Kissinger decided to stall the U.S. military resupply to Israel, thus gaining leverage over Jerusalem and earning the gratitude of Cairo. Days later, as Israeli troops surrounded the remnants of the Egyptian army, Nixon and Kissinger pressed Israel not to plunge the final knife. Israel complied with this request, thus elevating Washington to the status of Egypt's savior. As for Israel, it had no alternative but to stay close to the United States, its only ally. Nixon and Kissinger had successfully maneuvered to establish a psychological balance between the parties. For the first time, an Arab state had not been totally routed by Israel and had been able to retain enough pride to negotiate. For their part, the Israelis were sobered by their early setbacks and were readier to negotiate as well. The Americans used this two-way leverage to mediate Israel's pullout from Egypt's Sinai Desert and ultimately to produce a peace treaty between these two sworn enemies. Neither Israel nor Egypt would have entrusted this role to any other nation or group of nations, at the time or since, and, even if they had, none could have accomplished anything.

The third ring of the diplomatic circus was in Asia itself. Nixon and Kissinger perceived immediately that the more China benefited from the Vietnam War, the more Asian nations would desire American resurgence and protection—and that this desire was the key to U.S. power in the region. To Asia, the United States was the only nation capable of blocking future Chinese power, and being far away, the only nation Asians could trust to do so without taking undue advantage of them. Most Asian nations did not want the United States to be the loser and to appear powerless. Nixon and Kissinger leveraged these fears and desires as they reaffirmed U.S. military commitments in the region and increased military aid to friends there. Four years after U.S. Marine helicopters lifted the last of the Americans off the rooftop of the American embassy in Saigon, few disputed the fact that the United States' position in Asia was perhaps stronger than ever.

The third example of U.S. strategic brilliance occurred as the Soviet Union began to implode, and when the first president Bush, Secretary of State James Baker, and National Security Adviser Brent Scowcroft discerned that the most effective and least dangerous way to lay the groundwork for the Soviet leader, Gorbachev, to dismantle his Eastern European empire and the Soviet Union itself was to burnish his luster and his power. They made him a world hero and thus gave him the power to write the epitaph for his own nation and empire.

As Soviet troops lurched toward disaster in Afghanistan in the late 1980s, Moscow began to lose its grip on Eastern Europe, and later, it lost its grip on the various republics of the Soviet Union. Cold Warriors urged Bush, Baker, and Scowcroft to press their advantage to the fullest and convert the Soviet empire into a Western-style democracy. In other words, the triumphalists demanded that the United States not let this moment of clear superiority slip away. But Bush, Baker, and Scowcroft rightly saw that they could achieve their main objective—the dissolution of the Soviet sphere of influence and, in a few more years, of the Soviet Union itself—without backing the Soviets into a corner where they might dig in their heels or even strike back. The heart of Soviet power—its military forces—remained very much intact at this point, and the Communist Party bosses retained internal control. The party and the army were still in a position to assert themselves and snuff out opposition both in Eastern Europe and within the republics of the Soviet Union. It would have been bloody, but they could have done it, and the United States could not have done very much to thwart it.

Instead, Bush, Baker, and Scowcroft used U.S. power to push and guide Gorbachev, but they did this behind the scenes. In public, it appeared that the United States and the Soviet Union were shaping the end of the Cold War together. Bush made Gorbachev look like a world leader and a hero rather than a loser and, in this way, made him less vulnerable to Soviet reactionaries. And so, piece by piece, Gorbachev allowed the peaceful dissolution of an empire—and the reunification of Germany—without the firing of a single shot. It was a strategy of cooperative diplomacy on an unprecedented scale.

IN THE LAST TWENTY years, the majority of efforts to create a new overall U.S. strategy have run the gamut from inadequate to self-destructive. But there have been a few promising moments as well.

The members of George H. W. Bush's brain trust were well aware they needed something more than spontaneous ad hoc diplomacy to cope with a world brimming with unknowns and ever-increasing limits on American power. They ran out of time when Bush lost the presidency to Bill Clinton in 1992, but it was plain where they were heading: sensitive bilateral diplomacy with Moscow; coalition-based security as exemplified by the large number of allies they assembled in the First Gulf War against Saddam, achieved with the added bonus of the UN's blessing; and finally, the multilateral peace process for the Middle East they had launched in Madrid in 1991.

The constant ingredients in their unstated strategy were setting achievable goals (although they didn't get the Middle East quite right), clear and strong American leadership, a willingness to take the views of others into account, and a strong preference for joint action with other key states. They never swerved from their fundamental judgment that they could not accomplish any of their important goals alone. Bush was heading in all the right directions with his new post–Cold War strategy.

Clinton didn't so much reject these ideas as appear to think he didn't need a foreign policy at all. With the Soviet Union gone and China still far behind, Clinton and his foreign policy team didn't think America faced any serious threats. America, his actions suggested, could afford to concentrate on itself for the first time since 1941. A foreign policy would only distract from domestic priorities, and so the less foreign policy and the fewer foreign entanglements, the better.

To be sure, Clinton had his international accomplishments, some quite impressive: the eventual settlement of Bosnia at Dayton, concluding negotiations for the withdrawal of nuclear weapons from several former Soviet republics and their return to Russian soil and control, freezing North Korea's nuclear program, normalizing relations with Vietnam, helping the United Kingdom toward a peace in Northern Ireland, and acting to quell the financial crises in Mexico and the Far East.

But an overall strategy just wasn't there, aside from his continuing faith in the healing powers of globalization. Thus Clinton's successes appear to be isolated events that didn't lead anywhere or reinforce American power. Clinton's supporters and his former officials don't like hearing this, but leaders around the world faulted them severely both for their passivity in many situations and for their complete lack of any overall strategy.

George W. Bush rode into town with guns blazing against Clinton's foreign policy: no nation-building and a focus on Russia and China. At the outset, Bush's team rarely if ever mentioned terrorism or Iraq, although we now know that Iraq was already very much on their minds. Many Americans forgave the confusion over policy because they had confidence in Bush's highly experienced senior team of advisers, including Dick Cheney, Colin Powell, and Donald Rumsfeld.

The slate was wiped clean after 9/11. Bush immediately designated terrorism, Afghanistan, and Iraq as the principal foes, replacing China. Bush swore to destroy these foes—with the help of other nations if possible, but alone if necessary, and by military force. It was to be a crusade featuring U.S. military superiority. Bush took the unusual step of announcing that the United States reserved the right to strike other countries first and thus launch preventive wars, regardless of whether a direct attack on the United States was imminent. With domestic approval, he dispatched U.S. troops against the Taliban and al-Qaeda in Afghanistan. He then attacked Iraq, with relatively few allies, and threatened attacks on the two other suspected emerging nuclear weapons states, Iran and North Korea. There was no shortage of strategic clarity: America's international power would grow out of the barrels of American guns. To be generous, it was a catastrophe.

Even as this turmoil consumed his administration, Bush did a number of things that showed he had not lost all his senses. His administration and the United Kingdom achieved one of the most far-reaching and important agreements of the post–Cold War era: a pact that would grant Libya reentry into the world community and global markets in return for its renunciation of weapons of mass destruction and of terrorism, and its acceptance of full-scale inspections to ensure compliance. Bush also launched a farsighted and massive aid program to combat AIDS in Africa, which brought him widespread support there. Very belatedly, he led the multilateral coalition convened by China to negotiate steps toward an end to North Korea's nuclear programs, with appropriate inspections, in return for economic and diplomatic concessions. On these occasions where Bush worked well and compromised with others, the results were more promising—and most of all, indicative of what a viable strategy could be, even if talks with Pyongyang never reached fruition.

IF WASHINGTON WASN'T COMING to grips with a new strategy for the twenty-first century, other countries were, China winning the gold medal for the best strategy among them. Beijing's objective has been clear and simple: to maintain the rule of the Communist Party and to ensure it will not, and cannot, be challenged, internally or externally. It sounds like the old Soviet strategy, but it actually differs in three critical ways. First, China sees a relatively open and growing economy as the best way to maintain domestic control, whereas Moscow maintained internal control by force and intimidation with virtually no economic benefits to the general public. China has relaxed economic controls to promote expansion, but has so far defied the old adage that opening the economic floodgates inevitably weakens political control. And Beijing does worry about opposition because of widespread domestic unhappiness with the government's corruption and the vast economic inequality among China's population.

Second, whereas the Soviet Union's strategy centered on military power, China's revolves around economic power. Moscow's power lay in its military might and its cleverness in inciting wars of liberation in the southern hemisphere to embroil the United States and distract American power. While China increases its military might, it resorts to using little or no military power beyond its own borders, and in fact is highly critical of Washington for doing so. Instead of featuring its arms or ideology, Beijing's major use of its power tool has been its economy—as a buyer, a seller, and an investor.

Third, and again unlike the Soviet Union, China wants to look like "one of the boys," and not be too visible throwing its weight and power around. But at the same time, it expects to be treated like a great power, which it has become. In a sense, China acts the part of the un-American great power, making few demands on others, rarely scolding them, and always stressing multilateralism. It has improved relations with its neighbors, easing territorial disputes and promoting areas of free trade. It seeks a peaceful neighborhood in order to focus on considerable domestic challenges. It has not sent combat troops abroad since its unsuccessful invasion of Vietnam in 1979. Nor does it threaten to use force, except when Taiwanese leaders occasionally raise the subject of declaring independence from China.

Of course, China's strategic puzzle is far easier to solve than America's. Beijing's leaders worry about one thing—retaining internal power, mainly through a strong economy and accomplishing that goal through expanding trade and securing access to oil and other resources. They also rely on political repression and an appeal to nationalism. The 2008 Olympics offered dramatic examples of both.

Unlike Chinese leaders, U.S. presidents worry about everything. By contrast, China doesn't use its power to solve global problems; rather, it exploits the problems of others, especially Washington's, as when it plays off American assertiveness and unilateralism and acts the part of major power broker between "victim" states and the United States.

Because China's economic power is more formidable than its military or diplomatic power, it lets its economic power do most of its diplomatic work. And unlike Washington, China uses its economic power principally for economic purposes rather than to extract political concessions from others.

Like China, present-day Russia is clear about its strategy and the source of its power. President Vladimir Putin and his successor, Dmitry Medvedev, have sought to reinstall traditional Russian authoritarianism in Moscow, to destroy or diminish political opposition, to restore internal "stability," and to crown these goals by reclaiming Russia's role as a great power. Whereas the old Soviet Union had minuscule economic power and substantial military power, today's Russia has modest military power and growing economic power based on oil and gas riches. Russia is the world's largest producer of natural gas and the second-largest oil producer. Russia has the capacity to threaten or use force on its borders against weak neighbors, as it did in Georgia in August 2008.

Russia's strategy today rests on four sources of power: energy wealth, specialized sales abroad, its UN veto, and military superiority on its borders. First and foremost, Russia is accumulating its national wealth through oil and gas exports, and utilizing the profits to acquire foreign businesses, especially in Europe, mainly in the energy field. Unlike Beijing, Moscow blatantly throws around economic power, making threats to raise oil prices and reduce the availability of oil to neighbors. By and large, Moscow had been getting away with it until the world economic turndown of 2008. Nor has Russia been shy about leveraging energy for political ends. The nation is reconstituting its power. Second, Russia sells much sought-after nuclear technology and conventional arms to countries like Iran. The Russians have consistently ranked high among the world's exporters of arms and nuclear technology. Third, Russia has a permanent seat on the UN Security Council and the veto power that goes with it. Since so many nations turn to the Security Council, their need to avoid a Russian veto gives Moscow wide-ranging diplomatic power. Fourth—and this became clear after Russia's invasion of Georgia in 2008—the nation's leaders would remind the world that they held the military upper hand over adjoining states. With this strategy, Moscow's voice is increasingly heard and heeded on the world stage.

In sum, China and Russia are major powers to which others pay attention and which others try to placate, but they are not in any sense world leaders.

Last but not least are the Europeans and their chosen instrument for international power, the European Union. The Europeans have an overriding interest in a level of economic growth sufficient to sustain their very heavy domestic welfare and social support systems. The enormous costs of these programs have led the European nations to make deep cuts in their military spending and to focus their strategy on economics and diplomacy. The source of their power is their enormous internal market, now the world's largest. But they don't get the fullest potential benefit from this, because they have failed to unify on foreign policy or on security strategy.

Since individually and collectively they are a shadow of their former military selves, their strategy focuses on protecting their economic interests and on diplomacy, namely, on representing themselves as the guardians of international institutions and international law, international treaties, the United Nations, and multilateralism. They have maneuvered themselves into a leadership position as the champion of these organizations and concepts for weaker nations. By these means, the key nations of Europe remain players in the most important diplomatic arenas, but they are not leaders.

As for the United States, President Clinton and Secretary of State Madeleine Albright once came close to hitting the rhetorical mark on strategy. On December 5, 1996, on the occasion of her nomination as secretary of state, Albright elaborated on the idea of the United States as the world's "indispensable nation" that Clinton had just broached: "As the history of this century and the story of my life bear witness," she said, "the United States is, as the president has said, truly the world's indispensable nation. It is our shared task, with the help of friends from around the globe and of God, to uphold this proud standard in the years immediately ahead and into the next century." Then again in February 1998, in a television interview, she returned to this theme: "If we have to use force, it is because we are America. We are the indispensable nation. We stand tall. We see further into the future."

On the one hand, it was only a phrase—like Bush and Scowcroft's phrase "new world order"—but, on the other, Clinton and Albright were heading in the right direction, toward the idea that the United States has a unique role in international affairs. But neither of them ever described the precise nature of that role or exactly how or why the United States was "indispensable." At one point, Albright implied that the American role was that of a military enforcer. Then, she referred to "our shared task, with the help of friends from around the world." But she did not elaborate upon their role or ours. Nonetheless, Albright went on to become a constructive voice in trying to fashion the U.S. role in world affairs.

It is now almost a cliché when American leaders and policy experts refer to the United States as the indispensable nation. Some go farther and say that Washington must cooperate with others, "listen" to them, and "take their views into account." No one, however, has gone far enough and stated the obvious corollary, namely, that those with whom we should cooperate are also indispensable to us, and equally so.

So, there, hanging from the ceiling, was a blazing crystal chandelier, the central operating principle of the power pyramid: the principle of mutual indispensability. The principle expresses the essence of international power in the twenty-first century—that as powerful as the United States is, it can't succeed in solving or managing a major problem without the cooperation of other major countries (such as China, India, Russia, Germany, and the United Kingdom) as full partners; and that those increasingly powerful countries can't succeed in solving key problems without America. We swim together or sink apart. That is now beyond argument. Even Bush came around to joining and, ultimately, even leading coalitions to stop the North Koreans' and Iranians' nuclear efforts. But while acceptance of mutual indispensability is the beginning of wisdom on exercising international power, it is not the end.

It is equally essential to understand the roles that must be played by leader and partner. Brent Scowcroft expressed this point well in 2007, saying that "no other power or bloc . . . has demonstrated . . . the ability to mobilize the world community to undertake the great projects of the day. We, the United States, act as the catalyst. . . . The world is not susceptible to U.S. domination—but without U.S. leadership not much can be achieved."

The United States is the only country that can lead regarding major world problems, for several reasons: It has much greater economic and military capability than other major powers singly or together. Most countries distrust the United States far less than they distrust one another. And most countries in every region of the world continue to see the United States as the source of balance in their neighborhoods as well as their protector against potential regional threats. These are the unique strengths, attributes, and the international position that Washington brings to almost every international table—and from which it derives its unique leadership role.

But to understand *why* the United States is the sole leader doesn't explain *how* it can lead. One conclusion stands out clearly: in today's world, to lead is most certainly not to dominate. Other key states simply won't submit, nor do they have to. They have the power to resist, as do most states, large and small. Nor does leadership give Washington any special leverage to convince others that America understands their interests better than they do. To lead successfully, the United States must help solve problems—whether they involve trade, terrorism, the environment, humanitarian relief, or security conflicts. The only reason for other key countries to go along with American leadership is that they, too, need those problems solved and they recognize that without American leadership, the problems won't get solved.

America's power to lead thus boils down to the power to solve problems with the cooperation of others. In the end, that's the reason others follow us. Being the world's problem solver is the basis of U.S. leadership in the twenty-first century. Others can no longer be shoved around or exhorted to follow us on the basis of shared values that barely exist. Bush's second (and Obama's first) defense secretary, Robert Gates, put it very well: "Success will be less a matter of imposing one's will and more a function of shaping behavior—of friends, adversaries, and most importantly, the people in between."

But countries—Japan, Brazil, Russia, Germany, and others—will stay with the American lead only as long as they believe that the problem at hand is actually being solved or can't be solved by a coalition led by any other nation. Even then, they will stay the course with Washington only if they are satisfied that the coalition led by Washington is attending to their basic interests and making use of their perspectives on solving the problems.

For all the logic of mutual indispensability, the United States and The Eight have not been able to put it into regular practice. There are natural barriers to good sense. To begin with, other major nations simply don't see things as the United States does on many important issues. We differ even on the highest priorities. Fighting terrorism has a higher ranking in the United States than among the rest of The Eight, even if all for now are preoccupied with the economy. These nations agree on trying to stop the spread of weapons of mass destruction, but they strenuously dispute the urgency of the problem and how hard to twist the arms of transgressors. Many major countries disagree with us on the need, or even the utility, of pressing governments to modify or change their internal politics and values. Many countries believe that global warming is the top environmental priority, while Bush appeared to view it mostly as a liberal gimmick. Most nations still believe that the United States should be offering major concessions on trade, as in the past, but don't seem to appreciate that they are richer than ever while America is not as strong economically as it once was. These key nations, our essential coalition partners, don't show much more flexibility on these issues than we do. The upshot is that the world today has no coalitions and no solutions to major international problems.

To many Americans, mutual indispensability reeks of the vilest words in American politics: compromise and multilateralism. Conservatives, however, know that the United States needs to cooperate with other nations, and they address the problem in their fashion. John McCain asserted during the 2008 presidential campaign that "we have to be willing to be persuaded by others." This is as far as they will go, and it is striking that they will rarely if ever use the word "compromise."

To recognize that we need other nations and to accommodate their interests and perspectives is to commit the cardinal American

sin of compromise. In Washington politics, compromise equals capitulation, giving up good American interests for unworthy and even
evil foreign interests. An accusation of such sins can crush even a political saint or the cleverest bureaucrat. America is the only country
in the world where one has to explain that compromise is necessary
for cooperation, and that cooperation is essential for the resolution
of the majority of international problems. Of course, presidents can
deny that they're compromising and insist that they're just making
minor adjustments. Sometimes, they can even get away with this.
Presidents and their thoughtful allies can't engage in every fight over
every compromise; that's asking too much. But the way to avoid this
tiresome level of conflict and to give oneself the necessary flexibility
is to make the strongest sales pitch for the overall strategy of mutual
indispensability as the only way to solve problems. To put this the
other way around, if presidents don't sell the strategy, they will fail
more often than not. Thus compromise can be seen not as giving
away the family jewels, but as an essential ingredient in assembling
the power needed for success.

Some critics will assail mutual indispensability by saying that it
epitomizes the greatest sin of all—multilateralism. Only an idiot
would suggest abandoning America's "right" to unilateral action.
So, presidents should shout it from the rooftops—as long as they
don't try to practice it except in the direst circumstances. To make
mutual indispensability work, presidents have to escape from both
the concept and the word "multilateralism." It's too tarnished. Too
many critics have made that word sound like multilateralism for its
own sake or like tying the United States to a hopeless international
chain gang.

Mutual indispensability can and should be presented in another
way—as creating power coalitions of key countries to solve key
problems that could not be solved or managed by any other means.
This most certainly does not require that we beg any nation, let
alone dozens of nations, to join us in a love-in. It doesn't call for us
to go on bended knee to the United Nations, except when that suits
our purposes and when we can expect a positive result. It does mean

a working agreement with a few other key powers, specifically those that matter on the particular issue at hand and whose power must be added to ours to get the job done. We would not be debasing ourselves before others, but rather cooperating with others to do together what we could not do apart—tackling the issues that are fast becoming dangerous to all civilized nations.

Consider the consequences of having rejected this approach: Pakistan and North Korea have joined the nuclear club and Iran is now on the waiting list. Trade negotiations falter and protectionism looms. Terrorism continues unabated. Secretary Gates acknowledged in 2007 what everyone grasped, except for some of his colleagues in the administration: that "only a united front of nations will be able to exert enough pressure to make Iran abandon its nuclear aspirations." And to top it all off, the global environment is going to hell.

Presidents can protect American interests in these coalitions. As the leader everyone recognizes as such, an American president can set the policy direction for cooperation and, as often as not, shape the internal bargaining process and the terms of any compromise. Those are powerful assets. As any good bureaucrat knows, whoever gets to write the interagency paper gets more than his fair share of the credit for the outcome. But to be sure, presidents will have to make concessions and difficult compromises as well.

If presidents succeed with this approach, they can claim and will rightly receive their share of political laurels. If it doesn't produce results, they can always return to the practice of blaming others and asking for more time for the other side to concede. These excuses, however, wear thin after several years, compelling presidents to produce even flimsier excuses, which become the subject of ridicule both at home and abroad.

Not only the United States but other nations too must learn to appreciate the virtues of mutual indispensability. The burden of leadership, however, rests squarely on Washington.

In the end, mutual indispensability can succeed and coalitions can be forged only if the key countries all see two matters the same way: first, that the problem at hand must be solved, dealt with, or managed; and second, that only the coalition can best accomplish those goals. The ultimate incentive to cooperate is that without cooperation there is nothing but failure and danger.

Once upon a time in the Truman administration, the United States was known for doing its foreign policy business through international partnerships. It created and fully used international institutions. The Truman team realized that we received a very good return on investment through these institutions.

Most foreign policy and national security professionals today understand the necessity for these partnerships, and they're comfortable working with others. Foreign service officers and military officers, who have been on the line and have direct experience with the travails of accomplishing anything alone in the world, are particularly adept at this.

The resistance to the idea of mutual indispensability comes mainly from those who don't grasp the basic reality of power in the twenty-first century: Mutual indispensability is not an end in itself; it is only a means—and it is the only means of using American power effectively. Without coalitions, we will fail—no matter how powerful we are and no matter how firm our will is.

RULES FOR MAKING STRATEGY

Rule 1: Make the creation of a foreign policy and national security strategy your first order of business—and be personally and directly involved in the discussions. Without a good one, you'll always be chasing your tail.

More than anything else, it will help you to focus your substantial presidential powers and achieve victories that will afford you even greater power. More than anything else, it can save you from becoming swallowed up by second- and third-order issues.

Rule 2: Go through the whole strategy drill: determining achievable, not merely desirable, objectives; setting priorities; assessing everyone's strengths and weaknesses, especially your own;

deciding the sequence of your administration's moves, in particular analyzing what is the "first door" to open that will ease the opening of subsequent doors; and understanding the sources of your power to ensure that the means match the ends. A fool is someone who defines a problem in such a way that it can't be solved.

Rule 3: Make sure you get the bureaucracy, Congress, and the public to buy into your strategy at the outset. If you have to keep going back to the well to justify each individual action, you'll drown.

Professional military and foreign service officers will not be a problem, especially if the strategy calls for working with others, something they are skilled at doing. But do all you can to persuade your political appointees not to savage one another over the strategy, especially in public.

Legislators will leave you alone, most of the time, until you get into trouble. Remember that they love to be included in White House meetings, to be seen and heard at the White House.

The public will appreciate your efforts to tell them clearly and sensibly what you're trying to accomplish—both how and why. But don't expect to convert all the tree huggers, American Legionnaires, and foreign policy experts (they do have to make a living, after all.)

Rule 4: Your strategy must rest on the principle of mutual indispensability—with the United States as the indispensable leader and the other key nations as indispensable partners.

America's leadership power is based primarily on our capacity to galvanize coalitions to solve or manage major world problems. The partners bring the added strength the United States needs to make the coalition strong enough to surmount the inevitable resistance.

Alas, this approach requires mutual compromises, and that's no fun. In the last analysis, all parties have to recognize that if they really want the problems solved, there is no choice but to compromise and join the coalition. The reality is: Succeed together, fail apart.

You have tremendous bargaining advantages that allow you to set the directions and shape the terms of cooperation within the coalition. These pluses when added to the bargaining skills of the alliance can sustain you against inevitable know-nothing charges of compromise for its own sake, and multilateralism for its own sake.

Rule 5: You must repeatedly make clear that your strategy is to establish power coalitions of key states to solve problems, and not to practice multilateralism for its own sake. And no matter what happens, remind everyone that Washington always retains its right to act on its own.

Coalitions must include second-tier states (such as China, the United Kingdom, and India) and other lower-tier states only as necessary. You need to make clear that you have no intention of asking cooperation from a parade of states that would not be vital to the coalition's success. Do not cede control to the UN, but when possible, having UN approval doesn't hurt.

Finally, since nothing goes without saying: Good strategy is neither an inflexible doctrine nor a straitjacket that restrains action; good strategy is a good guide.

CHAPTER 6

INTELLIGENCE AND POWER

You presidents keep demanding of intelligence agencies what they can seldom produce—the innermost thoughts of foreign leaders and the status of concealed weapons programs. You rarely ask the spies for what they do best and what you need most—an understanding of foreign societies and their politics.

P resident Obama will put considerable treasure on the line to learn certain foreign secrets and peer into the future. Exactly where is Osama bin Laden? Are Iranian and North Korean leaders determined to pursue and hide their nuclear weapons programs or will they settle for peaceful nuclear energy? Will Iraqi leaders surmount their conflicts and govern in peace together? And before he transforms Afghanistan into another Iraq, Obama will ask whether the political future of President Hamid Karzai, our man in Kabul, can be salvaged.

Fairly or not, President Obama will be disappointed by the answers. He is bound to believe that his massive spying apparatus with its see-all, know-all technology can ferret out any information he wants. Again, he'll be disappointed. When his agencies don't produce clear-cut answers, and when he can't get the facts, his policy advisers and political enemies will fill the void with their own versions of the facts. If he becomes desperate or overly passionate, he might push them too far and actually get the answers he desires, as George W. Bush did with Saddam Hussein's nuclear program in 2003—but the answers will be incorrect.

Good information and foreknowledge can spell the difference between victory and defeat, as Niccolò Machiavelli explained long ago. In *The Prince* he wrote: "[F]or knowing afar off (which it is only given to a prudent man to do) the evils that are brewing, they are easily cured. But when, for want of such knowledge, they are allowed to grow so that every one can recognize them, there is no longer any remedy to be found." Knowledge is, indeed, power in two respects: First, whoever controls it usually controls the argument or debate; second, good knowledge allows for more precise and effective use of power, both to prevent harm and to do good.

For these reasons, the United States now devotes upwards of $50 billion yearly to a bewildering array of intelligence efforts to figure out the capabilities and intentions of other nations. Most of this budget buys satellite-produced imagery, which essentially involves taking pictures and other images from outer space; and communications intelligence, which essentially involves intercepting and decoding messages. A big chunk also goes to the military services for their tactical operational needs. A smaller part goes to the CIA for its analyses and on-the-ground espionage.

Because knowledge is power, presidents begin their day with briefings on the juiciest intelligence morsels. I read the written versions of these highly classified briefings when I was an assistant secretary of state. I found them to be less informative and interesting than the best daily newspapers, but almost always intelligent.

Over the years, these briefings and the more thorough National Intelligence Estimates delivered good value on matters such as whether Russia would increase oil prices, whether an Israeli prime minister was politically strong enough to make bargaining concessions, and the size and nature of China's military programs. But on most vital questions, the payoffs have been dismal: whether the Bay of Pigs invasion of 1961 could rally the Cubans against Castro; the danger of the shah of Iran's being overthrown in 1979; whether the Soviets would invade Afghanistan; and, later, whether the Soviet Union was on the verge of collapse. On these big-ticket items, the intelligence community either produced little or was dead wrong.

GIVEN THE POWER AT stake in battles over intelligence, presidents were not satisfied by such disappointing performances. Every one of them made major efforts to correct the deficiencies. Their first recourse was to throw more money at the problem. The threat of terrorism at home made spending on intelligence widely popular. Almost as popular in Washington is for presidents to try their hand at reorganizing the community, as the collective intelligence agencies are called. They move boxes around, create new layers on vast organizational charts, deconstruct other layers, strengthen the role of the White House, and make other well-intentioned efforts that rarely work. Then, they turn to hiring and firing CIA chiefs, then putting the old hands out to pasture and bringing in the new, only to then fire the new ones and bring back the old ones. To be kind, miracles have been lacking. The reasons for the community's shortcomings lie far beyond budget increases, reorganizations, hirings, and firings.

CONFRONTED WITH THIS STUBBORN reality, presidents often try to circumvent the system entirely by becoming their own CIA directors. Some will try to become their own private case officer by getting their information directly from their foreign counterparts, i.e., by personalizing intelligence. They figure that they've spent their lives sizing up people, and they know how to do that better than some CIA analyst or a senior fellow in a think tank. And they—not the analysts—are the ones with direct contact with foreign leaders and therefore are in a much better position to make judgments about where the other side's policy is heading, and to catch subtle nuances. They're almost always wrong about their independent judgment of foreign leaders, and they compound their errors by rarely passing on what they've heard in their private talks to trained analysts for evaluation.

Bill Clinton thought he had such a fix on Yasir Arafat that he gambled American prestige on the Palestinian leader's signing a far-reaching peace accord with Israel in 2000. Arafat refused to sign the pact. Similarly, George W. Bush parlayed with Vladimir Putin in 2001, and then announced: "I looked the man in the eye. I found him to be very straightforward and trustworthy. . . . I was able to get a sense of his soul." It seems Bush missed a thing or two in his peek into the Russian leader's soul. Several presidents were overly impressed by the former Pakistani president Pervez Musharraf's argument that if they didn't go along with his very modest moves against Muslim extremists, the extremists would take over his country.

President Kennedy introduced yet another technique for bypassing the community, one that others followed and felt served them well. He made the White House his personal de facto CIA and sent out his own analysts to peer into the abyss of Vietnam and report back to him directly, not to their respective bureaucracies. In 1963, JFK dispatched two leading experts on Vietnam with very different backgrounds to tour that country and tell him what was really going on there. Major General Victor "Brute" Krulak reported back to JFK that military action against the communist guerrillas was progressing at an impressive pace, that President Diem's South Vietnamese government enjoyed great popularity, and that his troublesome brother and secret police chief Ngo Dinh Nhu could be managed. Joseph Mendenhall, a career diplomat, countered that South Vietnam was in desperate shape, that Diem's regime was near collapse, and that the whole Ngo family had to go. Kennedy quipped: "You two did visit the same country, didn't you?"

The fateful issue before Kennedy was whether or not to deepen the U.S. commitment to prevent a communist takeover of South Vietnam. His decision would depend in good part on whether the experts could answer one question: Was this war winnable? If yes, he might well ramp up America's military and economic efforts. If no, he might consider retreating by stressing that Vietnam's fate rested in Vietnamese hands.

Hearing only more of the same contradictory accounts, Kennedy did what any self-respecting president would do. He hedged his bets in every direction, delayed making any firm decision, and played for time to further test the waters in Vietnam and at home. To know more, he unfortunately had to do more—but he would do only as little as possible so as not to lose. This is generally what presidents do when they can't find out what they want to find out.

THE DIRTY LITTLE SECRET is that it's very difficult to deliver the information presidents really want. No one, especially not an intelligence officer, can bear the thought of telling the president that the CIA can't answer his question today or even next year. But someone needs to explain to presidents the facts of life about intelligence.

First, it is difficult to predict political events or political positions. Indeed, why should our intelligence analysts be able to predict the future abroad when Americans can't begin to foretell what will happen in their own country? A couple of years ago, no one I knew predicted that Senator Barack Obama would be the Democrats' nominee for president, much less win the 2008 election. Second, it is even more difficult to get hard facts on the most dangerous matters, such as weapons of mass destruction (WMDs). If the information presidents seek is crucial to a foreign adversary, chances are that the adversary will take every possible measure to hide the facts, and almost invariably will succeed. Finally, to make matters worse, political pressure on intelligence experts, external or self-imposed, to conform their estimates to presidential policies and presidential preferences is inescapable, and this distorts evaluations. Knowledge is power, and every single power player in Washington will fight to define reality for the president—and that fight is well above the pay grade of intelligence analysts.

There's good reason why few experts in the United States or

elsewhere predicted the fall of the shah of Iran or the collapse of the Soviet Union. Great upheavals generally don't announce themselves; they erupt from the streets, from long-hidden and suppressed hatred that suddenly transcends the fear of police or military might. Novelists and screen writers would feel these tremors far more acutely than intelligence analysts assessing conventional evidence.

Adversaries will guard most zealously their real bargaining positions on critical issues. Experts will probably never stop debating whether North Korea is truly prepared to cash in its nuclear program in exchange for economic and political bribes. The same question roils experts regarding Iran. Alas, Washington has no spies on the inside in either regime who can relay such information. And even if it did have such a spy, the CIA probably wouldn't believe he was for real. The community has very rarely been able to recruit individuals capable of such high-level penetration. What's actually going on in foreign leaders' heads is revealed only by presidents acting and making proposals, which spark reactions and unearth the underlying aims (which, by the way, may or may not have been there at the outset of the bargaining process).

Similarly, adversaries do well at preventing America from spying into their most important secrets, especially their WMD programs. Conventional military capabilities are relatively simple to track, but countries are adept at hiding their plans for WMDs, and the precise locations where they are developing them. This should not come as a surprise to the White House, given America's own hideaways. Whatever Washington seeks most, the target will do its best to conceal. Adversaries know well when U.S. presidents have designs on their family jewels, and they have a nearly unblemished record of denying such information to Washington.

During the Cold War, the United States ringed the Soviet Union with listening posts and spy satellites and still couldn't track many of its secret programs. In 1998, Washington was clueless about India's plans to explode a nuclear device, despite nearly constant surveillance. The Indian military knew about U.S. satellite and listening capabilities, and essentially rendered them blind and deaf. Thus

Washington didn't get the opportunity to press New Delhi not to test. Nor did Washington ever have a high-confidence fix on Saddam's WMDs; nor, to this day, does it have a fix on the precise state of nuclear developments in Iran or North Korea. It could only draw deductive conclusions about their capabilities.

Presidents, senators, and many others compound the already serious problems of gathering reliable intelligence by the political pressures they impose on the community. Each side in policy debates is eager for the community to agree with its assessment. Professional analysts are well aware that they will be targeted for saying "the wrong thing." Thus, they often find themselves driven to claim that they have clear facts when they have little or no evidence, as they did with Iraq; or to exaggerate what they know, as they often did in assessments of Soviet military strength. Presidents, their aides, and politicians are never satisfied with the mush of reality and uncertainty, or with mere trends and tendencies.

The most common form this problem takes is the pressure intelligence officers feel to conform their judgments to official policy. Intelligence officers aren't ordered to do this, but they feel its weight on their analyses, and their bosses feel it even more. The Pentagon, the most powerful bureaucracy in the federal government, set up its own full-blown intelligence shop under Defense Secretary Donald Rumsfeld to send the CIA the message that it had better fall into step on Iraq or jeopardize its own relevance. Vice President Dick Cheney, the hawk of hawks on Iraq, visited with CIA officers in the run-up to the war to "listen" to their views—and shockingly heard nothing to contradict his own. During the Cold War, the hawkish Committee on the Present Danger terrified CIA analysts who doubted Soviet military superiority, accusing them of being communist dupes or just plain dopes. These stories linger and sting every analyst's memory.

This is not to say that all, most, or even a majority of intelligence is simply a regurgitation of political dictates. It is not. The point is to emphasize that once the president and his top advisers have taken a clear public stance on a major policy problem such as Vietnam,

Soviet military power, or Iraq, the intelligence community in its collective judgment rarely contradicts that policy with unwelcome facts or opinions—except, of course, in the form of anonymous news leaks. Nor is this hesitancy peculiar to frightened intelligence officers. I cannot count the times I've waited with colleagues outside the Oval Office or the offices of the secretaries of state and defense and heard one of them rant about how he would tell the boss some unpleasant truth, only to be ushered into the office with him and see him smile with total docility once in the presence of the great one. On hot-button issues, there is no escaping the politicization of intelligence.

A rare and odd exception was the 2007 National Intelligence Estimate stating that Iran had suspended its nuclear weapons program in 2003. But this, too, was a highly political document. Although the estimate credited Iran with continuing other related nuclear efforts, agency leaders clearly understood the likely headline: Iran's nuclear threat was not nearly as imminent as Bush or Cheney had insisted. This estimate read more like a declaration of independence than an analysis.

Intelligence is a major battlefield on which policy battles are fought, and a devastating weapon in wars over policy. When hawks convinced leaders in Washington that Moscow had military superiority, they won the battle to spend more on America's defense and to resist negotiating "from a position of weakness." If hard-liners persuade Washington's leaders today that President Ahmadinejad of Iran is a new Hitler, hell-bent on the destruction of Israel and the United States, they will win the debate on taking the strongest actions possible to overthrow him. Whoever controls reality tends to control the policy debate. These policy debates are won not by green-eyeshade analysts, but rather by legendary political hit men.

OF COURSE, PRESIDENTS WILL continue to vigorously and publicly seek the very facts and the insights that will forever elude even the most diligent of intelligence analysts. If they didn't do this, their political enemies would clobber them for not caring about the facts. They would be defenseless against the charge of wasting money on intelligence they disregarded. To say that the United States couldn't do something would be, well, un-American.

But as presidents do what they must, they can also privately focus America's intelligence efforts on what the community can deliver and deliver well—and help themselves carry out a viable strategy for the twenty-first century. Such a strategy would be based upon the principles of mutual indispensability and of building power coalitions with other countries to solve global problems. That strategy requires precisely the kind of information the intelligence community is good at—understanding the societies and politics of other nations. To create coalitions and take proper advantage of the American position of global leadership means that presidents must have genuine command of how other countries work, their leaders' profiles, and how to influence them. This is where the community's comparative advantage lies.

The community has expertise in this critical area of domestic politics of other nations, and all foreign policy is rooted in domestic politics writ large—national values, interests, personalities of leaders, culture, and power structures. From this brew comes a foreign policy, and to this brew U.S. policy must be addressed.

In early 2006, Secretary of State Condoleezza Rice stood on a podium at Georgetown University and announced the primacy of other countries' internal affairs in making judgments on U.S. foreign policy. "The greatest threats now emerge more within states than between them," she said. "The fundamental character of regimes now matters more than the international distribution of power," she added. These observations about threats and international power balances are eminently sensible. What's startling, however, is that Rice presented them as revelations—note the use of the word "now" in both sentences.

It follows, then, that presidents and their senior advisers must harness intelligence to this end. This means asking intelligence experts on a regular basis how policies under consideration can best be framed to influence foreign power structures. If the experts don't have this information, they can surely obtain it. Any reporter worth his salt can discover who counts, when, and where.

The questions to ask are obvious and include these: Will the policy help or hurt America's potential allies? Will it make adversaries stronger? Is it better to delay the policy announcement until there is more backing in the military or business community, or is the policy designed to build that support? Is there a way of stating the U.S. position publicly that will get the message across without offending the leadership? What other countries do the leaders listen to? The answers, in turn, will permit American leaders to fine-tune their power. And given how hard it is to wield power effectively, presidents need every advantage they can get.

Open sources, especially magazines, newspapers, television, and the Internet, can provide vital information about other countries, as can the many foreigners who work in and visit the United States. So can Americans who conduct business abroad. Presidents must not only encourage but direct the community to take full advantage of these untapped open sources of information.

To make their policies work, presidents must have fine-grained knowledge. They have to know, for example, whether offering economic benefits to Iran will strengthen the hands of pro-American businesses, or hurt them, or have little effect. They need to know how these benefits might be presented so that they could help pro-American interests. They need a good sense of how general and increasing economic and political discontent will affect Iran's parliament.

No foreign policy, no power move, is worth anything if it doesn't bite into the domestic politics and decision-making process of the target country. Otherwise, the policy is just posturing for domestic American politics or for the vague entity liberals mistakenly label "the international community." Policy too often serves just such narrow political or ideological purposes, and when it does, it doesn't advance American interests. Intelligence can help a lot on the strategy front.

. . . .

EVEN AFTER ASKING FOR the right kind of internal analyses, presidents should not become passive recipients of this information. They can view this knowledge through their own historical understanding or intellectual predisposition. And if the analysts take issue with these precepts, they ought to explain why. Doing so will lead to exchanges that provide the president with insights. Too often, the intelligence analyses that presidents read have been filtered and rewritten by political staffers, and presidents would be well advised to read them exactly as the intelligence analysts prepared them or receive a briefing directly from the analysts.

Meantime, history teaches a number of lessons for presidents and analysts to bear in mind.

First, politics predominates in the short and medium term; economics in the long haul. Political leaders will tend to their political needs before their economic needs. Economic trends and pressures take much more time to unfold and take hold than do political pressures. Thus, presidents should be wary of predictions about economic privations resulting in capitulation in the short or medium term, or of abrupt alterations in a country's foreign policy or bargaining position.

A number of Latin American leaders shun benefits from economic deals with Washington because such arrangements would unnerve their political base, which is their primary concern. Similarly, major powers have dangled economic rewards before Iran, North Korea, and Syria, but their leaders clearly coveted their internal control more than economic growth. Some might argue that the Libyan strongman Muammar Qaddafi was an exception, that he made great concessions in return for access to the world economy. But in fact he opened his economy mainly because he believed that doing so was essential to preserving his own domestic political power.

Second, transforming a country into a democratic free-market paradise, or even the local variation thereof, takes a long time—certainly more than an election cycle or a decade. The CIA's analysts and other government analysts are the least likely to forecast otherwise. This particular brand of optimism usually springs from policy advocates or politicians looking for excuses for what they want to do.

Third, nation-states rarely escape their own laws of history and gravity, and thus new leaders generally follow long-standing policies and seldom divert course fundamentally. Nation-states have a geography, a history, and a political culture that never disappears, and they move through certain stages of economic development with attendant social clashes and pressures. These underlying realities change only glacially, imperceptibly, as do the policies which are largely based on them. New leaders will tinker with changes in foreign policy, but usually make only minor adjustments over the short and medium term. European nations show such consistency, just as Soviet bosses largely stepped into the shoes of the czars (though their rhetoric was more adventurous), and the Chinese communists largely took the path of the emperors. Thus, presidents should greet with skepticism assertions of impending policy revolutions. It often takes a true political revolution, as in Iran in 1979 or the election of Hugo Chávez in Venezuela in 1999, to basically alter foreign policy, to go from pro-American to anti-American.

There has been only one country prone to fundamental changes in foreign policy without revolution—the United States. All it seems to take is an election. So, U.S. foreign policy has leaped from Nixon to Carter to Reagan to Bush to Clinton to George W. Bush. Chinese communist leaders did decide to open their economy in the 1970s, and their foreign policy evolved steadily from there. Either countries other than America don't face the quadrennial pressures to change policy, or their leaders lack the power of a U.S. president to make basic changes essentially on their own. Because Americans can switch policies so readily, they sometimes incorrectly ascribe similar capabilities to others.

Fourth, strong dictators always last much longer than American analysts tend to predict. That's because Americans believe that if things are really bad, people won't put up with them, but people do. Just take a gander at the following tenures, in no particular order:

Saddam Hussein, twenty-four years (1979–2003); Stalin, thirty-one years (1922–1953); Mao, twenty-seven years (1949–1976); Castro, forty-nine years (1959–2008); Kim Il Sung, forty-six years (1948–1994); Kim Jong Il, fifteen years and counting (1994–present); Mobutu Sese Seko of Zaire (Congo), thirty-two years (1965–1997); Hosni Mubarak, twenty-eight years and counting (1981–present); Saudi royal family, seventy-seven years and still counting (1932–present); Robert Mugabe of Zimbabwe, twenty-nine years and counting (1980–present); Slobodan Milosevic, eleven years (1989–2000). Strategies to oust these leaders or ease them out or get them to reform rarely fared well. None of this is to say that most dictators don't eventually succumb, but the process invariably takes a long time, and when they depart, it's almost always a surprise.

Fifth, moderates will act moderately and therefore cannot be counted on in a crisis or in fast-moving situations or turmoil. Moderates define themselves by their balanced judgment, tolerance of other views, caution, and willingness to compromise—all qualities generally prized in stable governments and societies. But these very qualities make for weak allies in crises, where the day usually goes to the passionate, ruthless, and most undemocratically organized.

The American tendency is to place hopes upon and reach out to the moderates because they share our values and temperament. Both Democrats and Republicans are much more comfortable with them. But in political cultures where moderates are most needed, they are in shortest supply and usually without power. American presidents have learned this only after hoping otherwise in Saudi Arabia, Iran, and Syria.

One of the most critical tasks for U.S. senior officials is to figure out how to help moderates in immoderate societies, without basing our policy on their likely success and without making them targets in their own countries for being pro-American. The U.S. intelligence experts should tell their superiors what I heard in the spring of 2008 from a dozen young Iranians, most of whom were very critical of the current regime in their country. They implored me to understand that whenever Washington tries to help reformers like themselves, the reformers are cast aside as American dupes. The community knows such things.

The key subject to ask the experts about is change. Is there a change in policy, or how can we encourage changes we desire? Most change comes slowly, and presidents should ask the analysts for justification when they claim to see otherwise.

EVERY COUNTRY HAS PROBLEMS obtaining good intelligence. Even with the best technological espionage and the best analysts, no one can escape the lenses and thus the distortions of his own culture and politics. As the Talmud says: "We do not see things as they are. We see things as we are." That difficulty is compounded in the United States because Americans' sense of themselves, and of being American, has historically been so strong and so inordinately isolating.

America has been cut off or remote from other societies because of its geography, relative newness as a country, wealth, comparative self-sufficiency, inward-looking culture, success, and power. America is a paradox, a nation of immigrants where so little seems to be known about the homelands of those immigrants. Foreign languages are brought to American shores and then quickly discarded and forgotten by the second generation. The CIA, the foreign service, and military intelligence and counterintelligence units have only handfuls of Arabic speakers, for example, although the United States has by some estimates at least 3.5 million residents of Arab descent.

Other major powers like Russia and China consider themselves special and have memorable histories to buttress their claims. Yet for all their pride, they and most other countries have striven to master foreign cultures and languages. They seem to have gotten the point about what it now takes to compete economically in the age of globalization and diplomatically in the era of pyramidal power. They know they must vastly increase their understanding of other societies. America as a country lags far behind and shows few signs of catching up.

Presidents don't have a lot of places to look for good information. They have the intelligence community, including experts in the State Department, and the community can draw on information from nongovernmental organizations, academe, and business. Not surprisingly, different experts within the community don't go out of their way to communicate with one another, and presidents have to push them in the right direction. Presidents also can save themselves considerable grief by observing the following rules.

RULES FOR INTELLIGENCE

Rule 1: Focus your intelligence experts on helping you to understand the power structures of other countries and how to influence them—because that information is essential to your basic strategy of building power coalitions.

To design such a strategy, you'll require fine-grained information. If you want to get serious about helping your friends and hurting your enemies abroad, you'll need specifics, not generalities. And you're in luck, because America's intelligence analysts have these skills and can get even better at them.

Rule 2: Push your spies to find out your adversaries' darkest secret thinking and most secret weapons programs, but don't expect good results. What you want most, the other side is best at hiding. And be prepared for others to fill in the blanks with "facts" you don't like.

Rule 3: You cannot escape fights, bloody ones sparked by worst-case-scenario artists, over the capabilities and intentions of our adversaries. Your harshest critics in news organizations and among policy experts and legislators will all have vested interests in painting you as weak and irresolute in the face of mounting dangers. Fight those fights if you believe the evidence does not justify worst-case conclusions. If you don't, you'll be pushed into actions ultimately regrettable to all except your enemies.

Rule 4: Don't waste a moment's time or a single dollar fancying that you can escape these intelligence shortcomings with "new" plans, schemes, and gimmicks. There is nothing new, and the record of these reforms and streamlining has never justified the brawling required to put them in place.

If tempted anyway, ask to be locked up until the passion subsides. Upon your return to sanity, ask the experts for more information related to making your policies work—and for different opinions. There is nothing more dangerous to presidential policymaking than unanimity among advisers.

Rule 5: Take consolation in one thing: As important as good intelligence is to successful policy, there are even more important factors. The Soviets had just about all the intelligence their hearts desired during the Cold War (such as double agents of the likes of Kim Philby, Aldrich Ames, and Robert Hanssen). These spies knew almost all our secrets and certainly helped keep Moscow in the power competition against a mightier America. Washington had no comparable spies. The Soviets won the intelligence war hands down—and lost the Cold War hands down. They had better intelligence, and we had a much better country.

CHAPTER 7

U.S. DOMESTIC POLITICS AND POWER

You dominate the politics of foreign policy much as Machiavelli's prince did almost 500 years ago. The problem is that foreign leaders don't see it that way, and think they can manipulate American politics against you, and thus they resist your power. You can do better at managing this game, but it's hard.

As the summer of 2008 rambled on, the world glimpsed a fine specimen of a foreign leader outmaneuvering an American president in American domestic. politics. As the presidential election approached, Prime Minister Nuri al-Maliki of Iraq and President George W. Bush were deadlocked over the future status of U.S. forces in Mesopotamia. Maliki could not accept anything that suggested an open-ended duration for the troops' presence. Bush was 100 percent against any hint of a timetable for a withdrawal. For weeks, they glowered at each other inconclusively—until just days before the visit to Baghdad of the then presumptive Democratic nominee, Barack Obama, who favored a timetable. At this point, surprisingly, Maliki told a German magazine that he now favored a "timeframe for a withdrawal . . . as soon as possible." After a few days of negotiating, Bush was trapped, and he and Maliki quickly settled on a "time horizon." Bush could not afford to be holier than the Iraqis themselves on keeping American forces in Iraq. He should have anticipated this trap and preempted it with language about the need for a withdrawal process, but American politics is an open business and presidents are vulnerable.

Foreign leaders delight in America's tumultuous and public battles over foreign policy and see them as opportunities for profitable meddling. This became clear to me in 1978, when I was the U.S. negotiator in the ill-fated talks with the Soviet Union to cut the transfers of conventional arms.

We were in Helsinki, Finland, and my counterpart was Lev Isakov Mendelevich, a courtly, old-fashioned, and shrewd career diplomat. He invited me to his embassy (which looked more like the palace of a viceroy) for a "four eyes" dinner (just the two of us). As the vodka flowed and the caviar kept coming, we quickly fell into a hilarious conversation about the entertaining vagaries of some American politicians of the day. With all this easy banter, I naively ventured to ask this wily Jewish survivor of Stalin's purges of the Soviet foreign ministry about his country's politics. "Leslie, I like you very much," he began, his hands moving into a pontificating position on his ample belly, "but you and all you Americans really don't know much about the inner workings of Soviet politics, and we're not going to tell you. It's our real advantage over you."

Frankly, I think the Soviets rarely got the better of us. Neither side gave very much during those torturous negotiations, but the Soviets thought their gaming of American politics gave them the upper hand, and so did most other government leaders. The leaders of Iran, China, and Pakistan, for example, surely make the same calculations about us as did the Soviets.

All our foreign friends and adversaries have access to U.S. executive branch agencies, Congress, our media, and our think tanks. They come here to tell their versions of negotiations, sometimes as accurate as, sometimes less accurate than, the administration's versions. They make unofficial proposals through these many open channels, where it costs them nothing to stir up the pot. Their lobbies in America—Taiwan's China lobby, Israel's various groups with their potent connections, Greece with Greek-Americans, and so forth—are not to be trifled with.

The United States has the world's most open political system. It is both blessed and cursed with more accessible paths to its government officials than any other society. Government secrets spill into the public arena, and virtually everything is discussed publicly, if not intelligently. Other countries romp in our playpen far beyond our capacity to meddle in their societies. They reach into Washington politics in hopes of getting Americans to compromise with each other before negotiating with them. They rejoice when Congress, the White House, and the Defense and State departments square off against one another.

Foreign leaders generally understand the iron law of international bargaining—that real international negotiations take place more within nations than between them. Far too often, presidents seem unaware of this iron law and neglect to protect themselves.

In practice, this iron law should benefit American presidents. They have enormous power and authority to do what they want in foreign and national security policy. They can keep the nation at war—as Truman did in Korea, Nixon in Vietnam, and Bush in Iraq—beyond the point of solid public support. They are free to frame U.S. negotiating positions without much interference from any other domestic authority, and most of the time, they do. Bush could and did resist entering serious talks to curb global warming, although a substantial majority of nations and of Americans felt otherwise. About the only area in which presidents are challenged effectively is trade negotiations, where the line between foreign and domestic interests is blurred. Presidents lose or give ground at home on foreign and defense affairs only after an extended period of costly failure—only when enveloped in perfect storms.

The problem for the president in conducting foreign policy arises from the fact that foreign leaders do not fully understand the magnitude of the president's prerogatives. They see opportunity, rather than presidential dominance, when they view America's political turmoil.

They forget that even badly wounded presidents can stay their courses, as Truman did on Korea over the wildly popular General Douglas MacArthur and as Bush did on Iraq in the face of his minuscule public approval ratings. Nonetheless, foreign misperceptions often lead foreign leaders to withhold compromises in hopes of flipping American politics against the president. The result is often stalemates in negotiations, as happens routinely on trade and security issues.

Presidential power slips abroad less because the president actually loses control over policy at home and more because he loses control of foreign perceptions of the domestic policy debate.

Presidents need to better manage these misperceptions of their vulnerabilities at home in order to protect their bargaining power overseas. To begin with, they have to remind foreign observers that the raucous mob of actors in American politics is not as influential as it appears.

PRESIDENTIAL POWER OVER THE making of foreign policy rests not only on formidable constitutional and political strengths, and on the deference given to the president as commander in chief, but also on the weakness of his competitors, whose limitations are generally underestimated, especially by themselves. The cast includes the think tanks, the nongovernmental organizations, the media, the lobbies, and the executive branch bureaucracy. Their weight is often enhanced by alliances with foreign governments.

To be sure, the players can bite as well as bark, and their clout at times can even be decisive, if they generate congressional action, that is, legislation. But they are mostly noisemakers, both the guardians of democracy and the wolves that prey upon it. The White House often characterizes them correctly as acting in their own self-interest and not the national interest, and as failing to offer viable policy alternatives. On occasion, however—and those occasions are of historic significance—they represent the finest checks and balances in the American democratic system.

Among those who periodically provide valuable information

and ideas should be counted the senior fellows and experts in think tanks or foreign policy research centers, as well as professors at universities. But the main function of senior fellows has become their writing of unmemorable op-ed articles in newspapers and their eager participation in unedifying shouting matches on cable news programs, mostly to the end of criticizing whatever administration holds power or whoever in that administration disagrees with them. They also constitute a corps of government officials-in-waiting, and this fact is not irrelevant to their unfavorable reviews of current policy. It's not their fault that even well-informed Americans are inadequately informed by serious journal articles and books.

The think tanks and many public policy schools at universities came to full influence during the 1970s and 1980s. Before, Washington housed only a few dozen or so such organizations dealing with international affairs; now, there are well over 1,000, depending on how many fellows constitute a tankful. They receive funding from big business, from individual donors, and from a kaleidoscope of ideological benefactors.

The most luminous of these research operations in Washington are the conservative Heritage Foundation; the neoconservative American Enterprise Institute; the moderate, centrist Council on Foreign Relations and Center for Strategic and International Studies; the liberal to moderate Brookings Institution; the liberal Carnegie Endowment for International Peace; and the relatively new centrist Center for a New American Policy. These and other such organizations nationwide employ thousands of senior fellows and assistants who, along with their confreres in public policy schools at universities, occupy a good deal of the public space in newspapers and airtime on television, certainly in professional foreign policy magazines, and increasingly on their own blogs. When a particular think tank's crowd holds the White House, its fellows praise presidential policies to the skies, while almost all the others condemn it with varying degrees of civility and substantive seriousness.

These experts hardly pose mortal threats to presidential policies and power, for one reason: Their commentary is usually far more analytical (the president's policies won't work) than prescriptive (here's how to fix them). They heat up arguments about policy, but seldom offer alternative policies around which a thoughtful opposition can grow. In addition, there are so many of these experts on virtually every issue that they tend to cancel out one another's views in the public arena. In sum, fellows and professors have added substantially to the general clamor and partisanship of the public debate, but lack the influence to tip the scales either for or against the incumbent in the White House.

THE SAME CAN BE said of the influence that nongovernmental organizations (NGOs) have on major policy issues. The NGOs emerged as major players in Washington during the 1970s and derived their influence from their genuine expertise on areas from human rights to refugees to the environment. They also had clout because, unlike the think tanks, they were action-oriented.

Many NGOs are dependent on governmental funds, and they have protectors in the executive branch and Congress who share their views. Many NGOs actually perform government-like functions, such as caring for refugees, and their governmental sponsors prefer these private endeavors to public ones. Usually, NGOs fly under the radar on major policy matters because each focuses on just one subject, and most avoid overall criticism of presidential policy.

Three notables among them are Human Rights Watch, the International Rescue Committee, and the Environmental Defense Fund.

Whereas there were merely a few hundred NGOs thirty years ago, there are thousands today, and in those numbers rests refuge for the president. They often cancel each other out. There are pro– and anti–global warming groups, for example. Once big business and other ideological interests saw the influence of NGOs, they created counter-groups to reflect their own interests. Many presidents learned of the existence of such counter–human rights groups only when they tried to crack down on Chinese human rights violations.

Even when most NGOs concerned with a single issue are unanimous, the president can choose to ignore them, as Bush did repeatedly with the global warming groups or as several presidents did regarding Chinese human rights violations. Like the denizens of think tanks, they don't have much clout in their own right. In order to have any impact, they must have the attention of the press and, more especially, help from Congress. Otherwise, presidents don't have to worry about them.

Presidents like to get NGOs on their side for broader policy issues, but they don't need the NGOs.

IT'S NOT THE NGOs or think tanks that preoccupy the White House; it's the media—almost as much as Congress, in fact. The power of today's media certainly exceeds that of the 1950s, when the Washington columnist Stewart Alsop sneered that the press had the "social standing of a dentist." To be sure, television anchors, star correspondents, and front-page reporters for prestigious newspapers can raise almost anyone on the phone and be seated next to the secretary of state's partner at dinner. In truth, my wife and I had better seats at sparkling Washington dinner parties when I was a correspondent or columnist for *The New York Times* than when I was an assistant secretary of state. But if telephone access and socializing are telltale signs of power, they're not real power. For the most part, the media can and do make life more difficult for most presidents, but the impact of the media on policy—the real test of power—is usually marginal. And when it's more than that, the reason is invariably that they are shining a light on a policy so ill conceived and so ailing that the administration is embarrassed.

The fact is that for all the media's control of the communications process among government players and between them and the public, officials do a better job of using journalists for their purposes than vice versa. In the overwhelming number of cases, the administration decides to put out a story—and does so with its own spin.

Truly enterprising stories that stem from independent, investigative reporting are rare.

The media add to this overall government advantage by building their daily coverage around daily executive branch briefings. In fact, they send their senior reporters to these set-piece events in the White House and the departments of State and Defense. Their reporting thus depends on what the government chooses to tell them in these deliberately uninformative briefings. Daily, the administration press officers call one another and decide what, if anything, to tell the press at the daily sessions. Usually, the line is something like "We are considering Iran's proposal," or "so-and-so's ideas are unworkable." Administration officials even decide who among them will speak to which reporters on the phone after the press conferences, and what the line will be. I know because I sat at both ends of these briefings. And these meaningless bits of trivia derived from this daily game often lead the news.

Many journalists suffer from a further disadvantage in covering the range of foreign, security, and international economic issues: They are not, as a group, overburdened with background and knowledge in these subjects. In part, this results because editors do not keep them on a beat long enough to learn the history of policy and its intricacies—a situation unlike the longevity practices in the 1960s and 1970s. Thus, most reporting scants substance and focuses on motives, personalities, and politics. I often annoy my former journalistic colleagues by telling them, "If you don't know substance, write about politics." Politics requires less expertise than substance. Everyone's a political expert and about on an equal footing. Politics is mainly opinion, the sources' and the reporters', and such is the state of the art in the twenty-first century.

Thus reporters rarely file stories on a thoughtful legislative speech or on a senior fellow's devastating critique of policy or a new policy idea. Sadly, most don't know what good is, and their editors often know less. It often appears that neither reporters nor their editors take the time to read major administration policy statements,

let alone tedious government documents that often contain vital information. Seymour Hersh of the old *New York Times* sat for hours
in the D.C. newsroom and read the *Congressional Record* and committee hearings, as he simultaneously talked on the phone. (Thank
heaven, in this respect, for the bloggers who do read this material.)

A lack of substantive backgrounds leads to another journalistic
problem: inattention to establishing the facts. This should be the
main business of the media, to help sort out the facts for citizens
who don't have the time or background to do it themselves. But
today, many stories are reduced to he said–she said, and people are
given little factual basis for judging policy arguments. The benefit
here, of course, goes to the biggest liars, since—without serious fact-
checking by the media—they often get away with it.

As for television news, it does headlines, and the cable news shows
have reduced the presentation of news to shouting contests, to opposing voices insisting on their own viewpoints and ignoring the
facts. Cable news makes little effort to perform the basic function of
journalism—to explain what we know and don't know about a situation. Cable news thus has reinforced an already hysterical partisanship in American politics. For instance, CNN spends more time on
the weather and children who have fallen into wells in Texas than
on possible pandemics. Presidents deserve their share of the blame
for this dreadful state of ignorance. They take few pains to improve
this process and educate the public. In a media world of spinning,
the White House is spinner in chief.

While the media do occasionally seize on a particular presidential policy, such as Iraq, make no mistake: reporters, editors, and
publishers do not want to tangle with the White House and its partisans. Yes, they'll shout questions at the White House press secretary, but seriously pursuing administration officials on facts and
policies embroils them in public fights. The media, which are very
thin-skinned and cannot bear criticism, just naturally shy away from
doing their jobs. They don't want to be accused of being partisan, of
being un-American, or of undermining our troops in the field.

While journalists will object to this judgment, I believe that they generally fear Republicans more than Democrats. Perhaps this is because most journalists are themselves secret Democrats. But it's certainly because devotees of the GOP play a rougher political game than the Democrats. When a reporter stings a Republican, the Republican will call the editor and complain. When a reporter stings a Democrat, the Democrat invites him to dinner. Democrats try to persuade; Republicans try to frighten.

If presidential policy appears to be going well, reporters keep their powder dry and don't mention the negatives they may have heard. It is mainly when policies hit the skids for an extended period that journalists muster the courage to provide the public with negative stories. At that point they pile on. And it is only in those moments that journalists really begin to exercise political power. Administrations usually respond by trying to cover up their mistakes, and this reaction gives news stories even greater bang.

The media cannot by themselves sustain an attack on presidential policy; they become too exposed. The media need events to keep the fire going. Reporting on Iraq makes this clear. From 2004 to 2007, news coverage was predominantly negative because most events in Iraq were predominantly negative. This coverage surely soured public opinion on the war and helped the Democrats regain control of Congress in 2006. But when American and Iraqi casualties declined toward the end of 2007, reporters carried that message, and the Iraq story often disappeared from the front pages, although there was still a great deal to report.

Those who want to demonstrate that the media are indeed all-powerful usually cite Walter Cronkite, the CBS anchor, who returned from assessing the Tet Offensive in 1968, stepped away from his anchor's neutrality, and told Americans, in effect, that the Vietnam War could not be won. Cronkite's standing was unprecedented and surely affected public opinion, but it still took almost seven additional years before the United States withdrew from Vietnam. For all the complaining about the power of the press, it's hard to find an example of the media ganging up on and besting the president on foreign policy. For the most part, they just give him daily annoyances.

. . . .

LOBBIES CAN GIVE THE president more than that. To comprehend their power, take a fanatical member of an NGO—a global warming diehard, for example—and cube his passion, and you will get a Cuban-American, an Israeli-American, the old Greek-American, and now the new Indian-American lobbyist. They have passion, and passion produces power.

My job at the State Department gave me the lead position on arms sales, and for a variety of strategic reasons, I pushed very hard to approve a wide range of arms sales to Israel. Hearing from their inside sources that I had been opposing these sales, the various groups in the Israel lobby struck hard at me. One of the many attacks was one I learned of from my own mother. She phoned from her nursing home in tears. "It says here in *The Jerusalem Post* that you're a cold Jew, and you won't help Israel."

In 1996, my wife, Judy, and I dined with Fidel Castro and a handful of his aides in Havana. Castro launched into his familiar rant against the Cuban American National Foundation, America's formidable anti-Castro lobby. "Jorge Mas Canosa [its leader] runs U.S. policy on Cuba; he's the one stopping trade and better Cuban-American relations against the wishes of most Americans," Castro thundered.

To his surprise, I conceded, "You're right. Interested and powerful minorities in my country, in almost every country, generally carry the day on their issues. Senior citizens control Medicare, the gun lobby on guns, teachers on education, American Jews on Israel, and yes, the Cuban American National Foundation on Cuba."

"So, you admit you don't have a democracy." Castro moved in for the kill: "Rule by minorities is no kind of democracy."

"But we in the United States have the right and opportunity to change the dominant minorities by democratic means ... unlike Cubans. And the president can still defeat these lobbies if he's prepared to fight," I countered. He who had never lost an argument was unimpressed.

The fact was that the Cuban American National Foundation *had* been running U.S. policy toward Cuba. Because of the passions of many Cuban exiles in America, their voting power in Florida and New Jersey, and their funding of political allies, few Democrats or Republicans wished to wrestle with them. But by the same token, no president thought that the issue of establishing better relations with Castro was worth a political brawl with the exiles. In the constellation of American national interests and threats to American vital interests, Castro's Cuba did not rank high in any White House. When the time comes and a president decides that the exiles have lost their clout, especially now that Castro is out of power, the president will establish new and friendly relations with Cuba. Until then, it is easier for occupants of the White House to let the exiles shape the policy. Cuba isn't that important.

However, it is also the case that various presidents have secretly kept in direct contact with high-level Cuban officials. Flying to Havana and holding meetings with Castro have been a favorite pastime of senators for decades. And when push came to shove, Bill Clinton did send Elián González back to Cuba, against the fervent wishes of Cuban-Americans.

As a rule, however, presidents don't want these unbridled passions to organize against them politically. Lobbies do make an enormous amount of noise and receive a great deal of attention. Thus, in general, presidents let these lobbies have their way—until a president decides that the issue at hand is sufficiently important to take a lobby on. When that moment comes, presidents do confront the lobbies and override their wishes.

The China lobby prevented Washington from having high-level contact with communist China from 1949 until Henry Kissinger's secret trip to Beijing in July 1971. At that point, Nixon and Kissinger concluded that they could maneuver China into their diplomacy as a counterweight to the Soviet Union, and that ties with China were of supreme importance to that goal. Kissinger met with the Chinese secretly to avoid counterpressure from the China lobby, and then publicly announced contacts and future ties with the mainland. With that one stroke, Nixon instantly halved the power of the China lobby. Of course, it greatly helped that he was a card-carrying conservative anticommunist. All the lobby could do from then on was to help make it impossible for Washington to abandon Taiwan to China, something no president would have done anyway.

As for the vaunted Israel lobby, it has been very successful at maintaining high levels of economic and military aid for Israel and at reinforcing America's support for Israel's security. But for all its political pull, the Israel lobby could not prevent U.S. arms sales to key Arab countries; nor could it stop administrations from endorsing, first privately and then publicly, a Palestinian state carved from Israeli-occupied territories in Gaza and the West Bank.

These lobbies pop up regularly. Most recently, Indian-American business leaders have joined to create a pro-India lobby and helped gain Congressional backing for a controversial nuclear agreement between India and the United States. American business leaders have effective lobbies to protect their economic interests with China. But if China starts to lob missiles at Taiwan or to confront Japan over oil in the East China Sea, those potent lobbies will immediately retreat.

Presidents generally give the lobbies a wide berth. They watch the power of lobbies wax and wane, and when it is waxing, presidents fear them. Fighting them is generally too costly—unless the White House feels that a vital national interest is at stake. When that happens, presidents do fight and win, and the lobbies lose.

LOBBIES, OF COURSE, FOCUS on politics and thus principally on Congress, but like everyone else in Washington, they pay great heed to the federal national security bureaucracy. This includes the departments of State and Defense and the various intelligence agencies. For the most part, these agencies serve presidents loyally, except when the White House challenges their bureaucratic interests. This is not a cozy environment for the White House, and it leads presidents to view their own career generals, diplomats, senior intelligence officers, and senior civil servants with suspicion. Foreign leaders are well aware of these delicate relationships, and therefore look to the American bureaucracy as an avenue for influence.

The bureaucracy is a beast unto itself with many varied unattractive parts. Most of those at the top of the bureaucracy—people in policy-level positions—are political appointees of the president, rather than career civil servants. The one exception is the military, although civilians (such as the secretary of the army) are formally in charge of military departments. Presumably, presidents trust their own appointees, but even here, they worry that appointees will be co-opted by the bureaucracy. Such presidential concerns sometimes even apply to their own secretaries of state and defense.

The bureaucracy can indeed burn presidents. For military officers, the trigger is invariably an adverse budget decision to terminate certain pet weapons systems or research and development programs, or perhaps a White House decision on cultural questions such as gays in uniform. At the first hint of such betrayals, the military appeals to its congressional allies, and it normally succeeds because legislators have an interest in manufacturing certain military widgets in their states or congressional districts. For diplomats or intelligence specialists, the boiling point tends to come either with decisions they believe to be profoundly wrong or with decisions about which they simply weren't seriously consulted. As with some military officers, their weapon of choice is a leak to the press. Many of these leaks find their way to the front pages. The stories hit the presidential solar plexus because the public is being told that the president's own experts, his handpicked professionals, disagree with the boss in the White House or were ignored or overridden for political reasons. These stories hurt because they're so simple.

Presidents and their aides also wring their hands about the close connections among American military officers, diplomats, and intelligence agents, and their foreign counterparts. They are in touch regularly—since that's their job—and funnel their views back and forth. Common outlooks evolve among them, which often set them at odds with their political masters both here and abroad. And in countries such as the United Kingdom, France, and Japan, the professional corps packs such wallop that leaders in these countries readily assume that their American counterparts are equally influential. They aren't, but misperceptions of their influence can cause delays in compromises from abroad.

For the most part, the professionals are, indeed, professional and loyal. They do what they are supposed to: namely, heed the president and his appointees. Complaints by the White House about bureaucratic disloyalty or unresponsiveness are largely misplaced. The fact is that generals and foreign service officers, like everyone else, want to get promoted, and usually find reasons to adjust their professional opinions and are able to support the president's policy. This was seen dramatically in 2006, when several retired generals announced their disagreement with Bush's Iraq policy, although they acknowledged that they had saluted the commander in chief when on active duty.

MORE THAN THE BUREAUCRACY, the lobbies, the media, and everyone else combined, Congress can stop the president in his tracks and force major changes in foreign policy—on those unusual occasions when it is roused from its slumber-inducing rhetoric. Until those rare moments, members of Congress do little more than aim a few darts of criticism followed by quiet acquiescence. Capitol Hill is an odd place that sometimes rises up like a giant, but mainly tries to avoid responsibility, especially on matters of foreign policy. Foreigners may be forgiven for not understanding when the sleeping congressional giant will awaken.

Legislators have sensible reasons for accepting presidential dominance in foreign affairs and security. They are uncomfortable challenging the president's constitutional authority as commander in chief. It's bad politics and bad form. There is also deference toward the executive branch in dealing with foreigners that stems from a proper respect for its expertise. And most legislators either rightly believe that the nation should speak with one voice overseas, or that they don't want to be accused of weakening this unified voice. In addition, it is all but impossible for members of Congress to agree among themselves on an alternative to presidential policy. So, legislators are almost always content to criticize and cajole the president, but not to take policy into their own hands. There are, after all, 100 senators and 435 representatives.

For Congress to threaten to or actually block the president from pursuing his course, or to mandate that he take another one, requires a highly unusual set of circumstances that rarely occur simultaneously.

Congress, of course, has handed presidents startling defeats from time to time. The benchmark was its rejection in 1919 of U.S. entry into Woodrow Wilson's League of Nations. Congress also tried to tie Franklin Delano Roosevelt's hands by passing neutrality laws, but Roosevelt found ways to circumvent legislative intent.

From 1939 on, Congress did not reject a major presidential foreign policy initiative until near the end of the Vietnam War. For a few years thereafter, it asserted legislative prerogatives and toughened the hearing process. Most importantly, it started cutting off President Nixon's military and aid options for Vietnam one at a time.

In 1973, Congress passed the War Powers Act over Nixon's veto. That act required the president to seek congressional approval for keeping U.S. troops in combat more than sixty to ninety days. It constituted an open and direct challenge to the president's powers as commander in chief by both Republicans and Democrats. Subsequent presidents, however, either worked their way around this resolution or ignored it.

The U.S. Senate also rebuffed President Carter in 1980, when it made clear that it would not ratify his strategic nuclear arms pact with the Soviet Union. This unusual step was motivated more by the Senate's frustration with Carter's failures in managing Soviet-American relations and the Iran hostage crisis than by problems with the treaty itself.

A Democratic Congress also opposed Reagan on arms control

issues. Public opinion polls showed a significant majority concerned that Reagan's tough talk about Moscow was increasing the risk of nuclear war. Congress went so far as to start legislating what missiles should be manufactured.

Congress dealt Reagan yet another blow on a major policy issue when it barred him from providing any kind of aid to the right-wing contras fighting the left-wing Sandinistas in Nicaragua. This action so outraged Reagan that he secretly authorized U.S. arms sales to Iran to gain the release of American hostages and to use the money gained from those sales to provide aid to the contras that Congress had forbidden. The upshot was the Iran-contra scandal, which Reagan weathered by semi-admitting his guilty role, although some of his subordinates faced jail sentences.

For presidents to lose these and other battles with Congress, almost everything has to go wrong—all at the same time and over a long period of time. Only then are legislators ready and able to foil the prince.

Below is a sampling of what had to go sufficiently wrong for Congress to rise up and defeat a president's foreign or national security policy:

First and foremost, the policy under attack was seen to be failing or to have failed, and to be taking too great a toll in lives and treasure, as in Vietnam, or posing great dangers, as seemed the case with Reagan's early anti-Soviet policy. The perceptions were widely shared in Congress, in the media, and among the oft-quoted public intellectuals. Usually, a majority of Americans also opposed the policies. Public opposition was particularly devastating, since it usually tended to give the president the benefit of the doubt.

Second, the failures were seen as calamities, not merely mistakes. In all likelihood, a majority of senators judged Carter's Panama Canal treaties to be a serious policy mistake and didn't want to approve them. But they didn't see the treaties as a calamity, and so Carter won, albeit by just one vote.

Third, the calamity had to be of long duration and be ongoing. It takes years for enough legislators to decide to strike down the president. Such opposition does not appear overnight. Congressional opposition takes many years to erupt, as with Vietnam, Reagan's policy toward Moscow, and the Second Iraq War. Congress drew its knives only after many voices had been raised in protest, and after the public and the elites had reached a thorough and all but unanimous conclusion. Let it not be forgotten that although there was a consensus against Bush's Iraq policy, after 2008 Congress never voted to halt it. In fact, one of the few pieces of legislation Congress passed against Bush administration policy on Iraq was one conceived by Senator Joseph Biden and myself on a plan for decentralizing power in Iraq to keep the country functioning and unified.

Fourth, the congressional knives must be plunged by members of both parties. The defeats are inflicted by a majority of one party and a minority of the other, but both must participate. And in most cases, Congress must be controlled by the rival party, not the president's. The only recent president whose own party controlled Congress but still rose against him was Carter.

Fifth, Congress must be assured of the president's general political debilitation before it finishes off the job. Specifically, this means that opinion polls must show a sharp drop in the president's popularity and a precipitous decline in his job approval ratings. Legislators did not actually strike down policies by Nixon, Carter, and Reagan until pollsters effectively assured them that the president couldn't strike back. Interestingly, Congress didn't vote down foreign policy initiatives by George W. Bush, despite his historically low public approval ratings.

Meeting all these conditions for overturning presidential policy on national security is no small matter. So, presidents don't often worry a lot about it. But on those rare occasions when Congress does rise up, presidents worry a great deal.

WHEN PRESIDENTS ADD UP all the potential domestic political threats to their foreign policy, it's enough to cause them to reach for the Maalox—even though they are well aware of the history of

White House dominance on these questions. It's the presidential dominance that is perennially misunderstood abroad, however, and it is not clear that any president in recent years has taken steps to shield himself better from foreign misperceptions and the attendant complications for U.S. international power.

This practical issue should not occasion wading into a grand philosophical debate and into the mazes of political accountability. Few seriously dispute the need for presidential dominance in national security. The White House is the only place that offers policy coherence in a messy political system and steady attention to the national interest. But presidents should not receive and probably will never receive more constitutional or political protection than they already have. There must always be room for the functioning of American democracy and for questioning presidential policies before wrong ones become irrevocably costly. The system has to give the president room to dominate but still allow room for him to be challenged. While presidents may have a monopoly on policy, coherent or not, they don't have a monopoly on wisdom. Many of Clinton's and Bush's foreign policies should have been taken to the woodshed and given far more public scrutiny both by the media and by Congress than they ever received.

The best and perhaps the only chance a president has to get a firmer grip on foreign policy debates, which will give him or her the necessary power abroad, is for presidents themselves to make the reasoning behind their foreign policy decisions clearer to the American people. The more debates head toward the gutter, the more irresponsible the charges, the less reliable the facts, and the less attention is paid to viable policy alternatives, the more vulnerable the president's position becomes. By becoming a party to these trash debates, a president diminishes his special stature and his power.

The troubles start with presidential elections. In almost every one of them since 1948, the candidates have grossly exaggerated threats and distorted their opponents' positions. Kennedy falsely claimed that the Soviets had vast superiority in nuclear missiles. When running for president, Barack Obama proposed a very short fixed deadline for withdrawing U.S. combat troops from Iraq, though few experts considered this a realistic proposal. In any event, there was not a serious discussion of Iraq or virtually any foreign policy issue in the 2008 campaign. It's usually the case during such campaigns that candidates exaggerate threats and oversimplify solutions.

Forgive me for singling out Iraq when the trees contain so much other low-hanging rotten fruit from which to choose. Bush claimed that the troop withdrawal plan offered by the Democrats would lead to certain defeat in Iraq and to terrible repercussions elsewhere. He and his minions charged the Democrats with doing the terrorists' work for them. The response of Bush and other Republicans was not a strategy; it was merely baiting the Democrats on terrorism. For their part, most Democrats kept talking as if it were at all feasible or desirable to pull out almost all U.S. troops in at most a year and a half. But it was simply not logistically possible to withdraw in that short a period without creating havoc in Iraq. Neither approach made much sense; the political system was deadlocked, allowing Bush to continue on his course. Few proposals linked withdrawals to an Iraqi political settlement, even though every counterinsurgency expert argued that this linkage was absolutely essential. And to show the absurdity of it all, Bush himself agreed in November 2008 to a three-year deadline for the withdrawal of all U.S. troops after having insisted during the campaign that such a deadline would be totally irresponsible. The Obama administration has continued to implement this timetable and agreement.

Such debates—and they are typical of debates on most critical security, trade, and environmental issues—don't protect the president's power or American security, and they certainly don't enhance American democracy. They feature false facts and false options.

While there's plenty of blame to go around, and so much that requires fixing, there is only one place to start—with the president. If the president demonstrates seriousness in public debate, if he and his aides take great care with the accuracy of their facts and policy alternatives and show civility toward opponents, there is a chance that others might follow. Even if others don't follow, the president retains command of the high ground.

The public expects more from the president than from other policymakers. The people expect that he will look after the nation's security, and it is all but impossible to persuade them otherwise. They're willing to support presidential policy adjustments and changes if things don't work. If Bush had reached across the aisle and put together a bipartisan and sustainable strategy for Iraq, the great majority of Americans would have honored him. He could have legitimized bipartisanship, regardless of what extremists said. And in the end, it is the president who has the most riding on success or failure because, more than anyone else's, it will be his success or failure.

The answer is not greater presidential assertiveness regarding the rights and prerogatives of the office. What is worthy of the country is the public sense of great presidential seriousness and fairness in conducting the public debate. To achieve this, there are a few good rules to observe.

RULES FOR PROTECTING PRESIDENTIAL
BARGAINING POWER ABROAD

Rule 1: Don't ignore or underrate the problem of foreign leaders' calculating that they can sway America's open political system against you and thus outbargain you. It's serious.

You've got to do a better job of letting these leaders know that you do dominate the politics of foreign policy by better managing the public debate over foreign policy.

Rule 2: Be serious about including as many key players as possible in your major policy decisions. That's the best way to get them invested in the policy, or at least more politically respectful of the policy. It takes a lot of time, but it's worth it.

This starts with your own bureaucracy—the professionals in the foreign service, the military, and the intelligence agencies. They'll add to your understanding, and including them will head off their invariably painful and embarrassing leaks to the press and Congress.

It also means including key legislators of both parties. And don't turn these into the usual "you listen/I talk" sessions. Don't just brief them and try to sell the policy; give them the opportunity to participate and argue. You might even make one or more of "them" top officials in your administration. Bipartisanship was always difficult to achieve—and is even more so today—but a little participation and bipartisanship goes a long way.

There is no reason not to include the NGO leaders, senior fellows, and professors as well. They may say something of value, and like everyone else, they do love to be consulted.

Rule 3: You must break the hold of television and particularly cable news on the public debate—or you will fall prey to the now common scenario of a news network announcing policy before you're ready to have it made public. This allows the least qualified among us to stand in instant judgment of your policies.

For a decade or more, your predecessors have felt compelled to respond to the 24/7 news cycle established mainly by the cable news networks. They gave answers before they knew the answers and well before having the necessary consultations.

More than any other force in the country, the cable news networks have taken control of the debate out of your hands and put it in the mouths of uninformed, highly partisan political commentators. Take back control: stop responding to them on their schedule and assert your own. Just instruct your press spokesman to say, "We're studying that, and we'll get back to you." Talk show hosts will vilify you, but most Americans will cheer.

Rule 4: Try to build a new political center of policy realists from moderate Democrats and Republicans. It should not be massive, but it can be a great help to you. This center would be your most reliable political base and would send a strong signal of your control to foreign leaders. You can do it.

You, more than anyone else in American politics, still represent our national interests. That's a big cudgel. Also, a decisive number of Americans are thoroughly disgusted with transparently vicious partisanship, and will welcome your seriousness.

Rule 5: Good policy is good politics, and your greatest political strength is to be seen as above petty politics.

When your predecessors took well-conceived actions and explained them well, they succeeded far more often than not. When they performed poorly abroad, they suffered at home. And their failures were largely the result of their pursuing short-term political gains or satisfying ideological partisans.

When you put partisanship aside, people immediately respect this. Members of the public will understand that you are their best advocate to act in their true interests. This is your greatest strength. It is what makes you special. You trash your own pedestal whenever you stoop to pure partisanship. And the average citizen, who may know little of substance, does begin to sense when you're serious and when you're not.

If you design policies principally to solve foreign policy problems more than to corner domestic political opponents, you will have the best chance of succeeding abroad and receiving credit at home. If you fashion policies mainly to vanquish domestic rivals or favor domestic allies, you are more likely to fail abroad and at home as well.

CHAPTER 8

MILITARY POWER

It's good to buy the biggest and best military sticks for all to see, and just in case. But use them very sparingly in this messy and sponge-like world of insurgencies and terrorists. Think more about power packages: the poised, posed, and credible sword, wrapped in diplomatic and economic power.

The post–Cold War world heralded an unprecedented era of peace among nations in modern times. It seemed so different from the one portrayed almost 500 years ago by Niccolò Machiavelli. "A prince should have no object, nor any other thought, nor take anything else as his art but that of war and its orders and discipline," he wrote, "for that is the only art which is of concern to one who commands." Princes, kings, prime ministers, and presidents didn't need much persuading about the virtue and necessity of war.

But no more, or so it seemed and so it was hoped, with the end of the Cold War. For the vanishing of the evil Soviet Union and the survival of the good United States as the sole superpower promised a Pax Americana. No one would dare tangle with the unmatchable military superiority of the United States.

Yet, in the less than two decades since Moscow's collapse, the United States has resorted to major armed action more than twenty times, compared with fourteen during the supposedly trigger-happy half century of the Cold War. And for all its military might, the United States did not stop India, Pakistan, Iran, or North Korea from developing dangerous nuclear capabilities. Nor has it dissuaded China from a yearly double-digit expansion of military spending, or halted the growling of the resurgent Russian bear. Nor did it win victories in Iraq and Afghanistan, despite America's imposing military efforts in both countries. Nor could Washington reverse the mounting and deadly threats from terrorists and religious extremists.

In certain respects, the world seems more unstable and perilous now than during the Cold War. Then, the superpowers probed each other and fought proxy wars, but stayed away from direct confrontation (except for the Cuban missile crisis), and most other countries felt relatively safe from attack. Today, volatile countries like North Korea and Pakistan possess nuclear weapons, and terrorists can and do strike devastating blows almost anywhere, making many a leader increasingly worried about his nation's vulnerability. American leaders certainly are.

Throughout the Cold War, Americans broke each other's bones and reputations over what constituted mortal threats to our vital interests; whether to fight or, if not, how otherwise to deal with such threats; when to fight and die over less than vital interests; when to help *others* fight and die instead; what not to fight for; and what means other than military force to employ to foil serious threats. Well into the first decade of the twenty-first century, Americans are still embroiled in the same arguments.

These debates won't ever be settled, even among friends. But we can certainly improve on our dismal record of thinking about military force and power. Time and again, my fellow national security experts and I have much too expertly and quickly proposed or endorsed dispatching American armed forces into battle. As often as not, most of us have come to regret it, rethink it, or at least pretend otherwise.

There are many questionable reasons behind this penchant for war, especially the career effects of looking weak, of being carried away by the same emotions as the public at large, and of the impulse to support the president. But there has also been an enormous amount of intellectual confusion in how we experts have thought about when and how to use force.

Most carelessly, in both public and private talk, politicians and experts tend to lump together the three related but distinct basic elements of the military machine: capability, force, and military power. Military capability consists of arms, fighters and all related personnel, equipment, and training. Military force is the actual physical act of using this capability in combat. Military power is the art and skill of employing both one's military assets and one's proven military prowess in war to threaten what other leaders value most: their lives, their power, and sometimes even their sacred honor.

In the twenty-first century, military power should take center stage in our thinking, and be the star of our national security policy, along with economic and diplomatic power. These three instruments of power constitute today's most effective power package—the best way to get our way without war. This power package worked with Iraq's Saddam Hussein, although he was afraid to let us know and thus look too weak. It worked with Muammar Qaddafi of Libya, who abandoned his weapons of mass destruction in return for economic and diplomatic benefits. For some time it even seemed to be having a good effect on Kim Jong Il, North Korea's quixotic leader. It really hasn't been tried on Iran. And if this power package fails, presidents can always draw and bloody their swords—and attract more wartime allies by having first taken the nonwar path.

Admittedly, power packages, for all their punch, allow for imperfect agreements. They entail compromises, which some people mistakenly equate with capitulation. Compromises do sometimes leave outcomes fuzzy, and in their wake, arguments continue. These shortcomings, however, have to be weighed against the alternative of war, where the costs are always very high—and where our side doesn't always win, either. And good compromises, welded to America's considerable power advantages, give us the chance to resolve outstanding issues on favorable terms in subsequent bargaining rounds.

The moral here is simple: American leaders need to think less about imposing their will through military force and much more about shaping events through the application of military, economic, and diplomatic power. That cautionary note hits hard and true when you examine the very threats and scenarios that have propelled America's leaders and military planners to make war instead.

THE THREATS THAT GENERATE war scenarios inside the Pentagon always start with ritualistic bows to big power menacing from China and Russia. Traditionally, the military services feel most comfortable with sizing their forces and developing capabilities to defeat the largest conventional threats, whether or not those threats are at all likely. To the Pentagon's credit, more and more planning time is now directed at three interrelated new threats: terrorism, the "king of threats" in the post-9/11 period; rogue states led by dictators determined to acquire WMDs; and rogue or failed states transferring those WMDs to terrorists.

Defense Secretary Robert Gates in 2008 moved these new threats to the center of U.S. military strategy. His new strategy requires the military services to focus their weapons purchases and training on irregular warfare, or what used to be called counterinsurgency. In other words, the services will buy fewer heavy tank and artillery units and place more emphasis on weapons designed to ferret out terrorists and fight them in the streets. There will also be less emphasis on short, all-out, and traditional combat, and more emphasis on the "long war" against terrorists, and on political and economic programs to diminish public support for extremists. This involves a massive dose of programs to diminish the environment that fosters extremism. To Gates, these nonmilitary efforts to help our friends govern effectively and develop their economies have the primary role in the new strategy, over force. As for the big potential threats from Russia and China, Gates calls for "collaborative and cooperative relationships" rather than confrontations. In the next years, we'll see whether the military and the politicians will go along with this new approach to buying military capability.

But American leaders have to do this with their eyes wide open. It means a commitment to defeating extremists and terrorists through nation building, an enterprise that may well do the job. The new strategy seems to shift the burden of responsibility for fighting the enemy onto the United States and away from the nations we are trying to assist. Before committing to such a course, our leader must be very confident that the peoples of the nation we're trying to help will fight harder for their own freedom than our armed forces will. We can never win in Afghanistan if we have to fight harder than the Afghans to defeat the Taliban, or prevail in Iraq if our sacrifice to defeat al-Qaeda must exceed that of the Iraqis. Nation-building enterprises can succeed only if based more on the will and capability of our friends than on our own.

Then there's also the question of whether our armed forces can maintain their capability to fight major conventional battles and win decisively and at the same time train and equip for nation building, including counterinsurgency. It sounds like it might be easy but it isn't. Nor can advocates of this dual approach sidestep the problem by asserting that the State Department should now develop the requisite capabilities for nation building. This would involve a massive transformation of a debilitated department and massive transfusion of personnel and training that wildly exceeds State Department attitudes and management skills.

MILITARY CAPABILITY IS PAID-for, acquired potential military force. It is the sum total of personnel, training, weapons development, weapons, intelligence, information systems, mobility, and sustainability plus the X factors: operational and organizational skills and leadership. What results is the ability to develop, deploy, and maintain firepower almost anywhere at the president's directive, sometimes for minutes, sometimes for years, whatever it takes to defeat the enemy on the field of battle.

The price tag for this in fiscal year 2010 was a requested defense budget of $664 billion, including the costs of the Afghan and Iraq wars. That's about equal to the total military expenditures of the rest of the world combined. Nothing like this disparity has occurred in modern times, and probably ever. Nor is U.S. spending likely to recede after the fighting stops in Iraq and Afghanistan. Future spending bills will stay at this level indefinitely because these wars have depleted stocks and have deferred investments, innovations, and further integration of operating systems; and because of new plans to expand the overall size of ground forces.

Over the last fifteen years, the tab reached $6 trillion, which has allowed the United States to construct the most potent military capability in history. For the foreseeable future, it is the only country that can fight wars like those in Iraq and Afghanistan for months, let alone for four or more years, far from its borders. No other country can move fully modernized strike aircraft carriers to a region and threaten mayhem. The three U.S. carrier strike groups at one point deployed near the Gulf could alone destroy every significant military and economic target in Iran. The United States alone can find enemies almost anywhere, when they get careless, with drones or spy satellites, and can fire missiles to destroy them in their cars or hideouts, as they watch reruns of *Seinfeld*. We alone can destroy almost any country's capital and drive its leaders from power with conventional firepower in weeks.

Military force is the actual use of military capabilities, the ultimate physical act. Its purpose is to punish, to signal the will and capacity to do even more harm. It can be utilized to destroy an opponent's own capabilities and will, to compel him to change undesirable policies, or to remove him and his government from power and install a friendlier regime.

The enterprise of gaming the actual use of military force invariably begins with the biggest threats—today, still Russia and China. They can deliver a mortal nuclear punch to American soil. In overall military capabilities they exceed all other nations, although they trail well behind America. Nonetheless, it's hard to find someone not confined by a straitjacket who believes Moscow or Beijing actually would launch such an attack or has any conceivable cause to do so. And if either of them did, the destruction all around would be so horrendous as to stretch even a neoconservative's notion of victory. No one should trouble about this mortal threat, though gifted military planners still draw the strange assignment of planning for it in today's Pentagon.

But what if Moscow or Beijing were to attack its neighbors? Would that also be a mortal threat to the United States, as some argued during the Cold War? One such hypothetical situation became actual in the summer of 2008 when Russia stormed and occupied two next-door provinces of Georgia and also moved its troops into Georgia proper. The response of the West was diplomatic and relatively mild, since its leverage was essentially political and economic, not military. There was some talk of accelerating Georgia's membership in NATO, along with Ukraine's. There was also a good deal of talk about the United States and NATO providing more arms to nations like Georgia and Ukraine, but no amount of arms would be sufficient to right the military balance in these border areas with Moscow.

Washington and the West are in a true military bind here. They can increase their rhetorical commitments to Russia's neighbors and Eastern Europe, send more arms, provide more military training, and even do joint military exercises. But the Russians might well make some military response that would demonstrate their superiority and Western weakness.

The leaders of NATO would do well to remember an ancient rule about power: the worst way to protect your power is to show your weakness. Western answers to these military problems will have to come almost entirely in the strategic and political realms. Western leaders will need both credible economic and political sticks and the carrot of prefiguring a new strategic relationship with a renewed major power.

If China made new and serious threats toward Taiwan and reinforced them by military deployments, and Taiwan had done nothing to provoke these actions, the United States would stand by Taipei at least up to a point. The most recent example of this was in 1996, when Beijing made warlike noises and deployments. Clinton dispatched two aircraft carrier battle groups to the area, with lots of talk of interposing them between the mainland and Taiwan. But Chinese leaders warned against this, and although the carriers hovered around Taiwan, they never actually went into the Taiwan Strait. In any event, the crisis soon died down. Obviously, both sides can do each other great damage in even the briefest military exchanges, if threats evolve into warfare. But Washington and Beijing both know full well that the United States does not have the capability to match Beijing's in sustained, intensive combat near China's borders. China has the clear advantage in this regard, and so the name of the game for Washington is to urge Taipei not to provoke Beijing and to rely on deterrence if tension mounts.

The threats and scenarios that detour military planners and politicians and impel them to seek solace in alcoholic beverages lie below Russia and China on the global pyramid of power—in the medium-tier countries such as Iraq, Afghanistan, and Iran, with the prospect of large numbers of troops engaged in small-unit actions against well-organized and dedicated foes who dwell mainly in the shadows. These are the Vietnams that loom in the twenty-first century. What drove presidents into Vietnam was not the importance of Vietnam in and of itself or any threat Vietnam itself posed to the United States. The threat derived from the domino theory. Americans believed that Moscow and Beijing were testing American will in Vietnam and that if they let this vital piece fall to communism, other nations would also fall in rapid succession. In the case of perceived threats from Iraq, Iran, and Afghanistan, they are seen to be dangerous in and of themselves. Afghanistan's government harbored terrorists who attacked the United States. Saddam Hussein looked as though he wanted to gobble up America's oil-producing allies in the region and seemed headed toward further aggrandizing of his power by developing weapons of mass destruction. To many, Iran appears to be a potential nuclear threat to Israel, an intimidating force against America's oil dependencies in the region, and a welcoming haven for terrorists.

In thinking about U.S. military action, these three states may look alike, but they are not. At the surface level, the United States has the military capability to attack the capitals of these countries and depose their governments. That's precisely what happened in Kabul and Baghdad, and it could certainly happen to Tehran as well (Iranian leaders must surely think about this).

But at the next stages of combat in these countries, the scenarios sharply diverge and turn muddy. Iraq started as a U.S. military triumph, then looked like a failure for more than four years, and then seemed to stabilize. President Bush did not deploy adequate forces, nor did our generals employ an effective counterinsurgency strategy. Once they began to take the necessary steps in 2007 and—far more important—when they invited political alliances with former Sunni enemies, the situation turned positive.

More critical still was that in 2008, the very Iraqi leader whom we installed and were even trying to ease out of office for his incompetence blossomed, totally unexpectedly, into a seemingly competent leader. This was Prime Minister Nuri al-Maliki, who then also managed to mitigate another major problem, namely the Shiite militias that had been wreaking havoc. Equally important, by 2008 the Sunnis and Shiites found it increasingly difficult to kill each other because ethnic cleansing had divided the groups into two areas of relative safety. How this will end as American troops are being withdrawn in 2009 remains uncertain. But in contemplating future Iraqi-type ventures, it's well to remember that the reason for going to war in the first place—Saddam's possession of nuclear weapons—proved false and that the direct and indirect costs of even a favorable outcome will probably exceed $1 trillion. If Saddam had been on the verge of attacking his neighbors on whom we depend for a major portion of our oil supply, and if he genuinely posed a nuclear threat, then perhaps the cost in lives and treasure would have been justified. Neither nightmare, however, turned out to be the reality.

The real reason for the continuation of the war in Iraq is that once the United States committed its prestige and credibility to war, its leaders felt they could not squander that (or their) credibility. In other words, we fight because we fought, and because Bush committed his own prestige and future reputation to the war.

In Afghanistan, the initial success of U.S. forces in routing the Taliban government and its al-Qaeda allies from Kabul devolved into what now appears to be an open-ended, dismal insurgency. The Taliban and company are coming back; the United States has been receiving only modest help from its allies in NATO, and even more troubling, its allies in Kabul are proving corrupt and inept. Democrats and Republicans alike now call for more American troops there, as well as more U.S. aid. They all seem to forget that no outside power has ever tamed Afghanistan.

What then is to be done with this country, one that might again become a serious threat to the United States if U.S. troops withdraw and the Taliban were to return to power? The answer is that there is no answer that rests primarily on U.S. or NATO military force. It's almost certain that neither Washington nor its allies will again commit massive forces to this fight. The Soviets did and were characteristically ruthless there, and they still departed with their tails between their legs. The United States will increase its troop levels and wage a more expensive war inconclusively for years to come. At the same time, everyone who knows Afghanistan well will be saying that a beneficial outcome for the United States depends on our Afghan allies themselves. We can give them military cover and train them, but they have to build a legitimate and effective government that the warlords and other Afghans will fight and die for. Even then, NATO forces will not be ready to counter effectively the Taliban's ace in the hole, its safe haven in Pakistan.

Iran presents equally daunting obstacles for military planners. The United States can utterly destroy Tehran and major economic assets with air strikes from stealth bombers and cruise missiles. The temptation to do so would be great. It would also be tempting to target Iranian oil fields and ports in retaliation for any terrorist attacks sponsored by Tehran. Military planners, however, certainly wouldn't recommend sending in sizable ground forces, almost regardless of the provocation. Hundreds of thousands would be needed if the mullahs and their allies were determined to fight. In other words, we have no real prospect of cementing an initial military victory by controlling Iranian territory.

The United States can also conduct preemptive surgical strikes against Iran's nuclear facilities, but success here is easier said than achieved. American intelligence is far from sure that it knows the exact locations of these targets and whether they can be destroyed by air attacks alone. And even if the targets are destroyed, they can be rebuilt. And while they're being rebuilt, Iran could hit back on American soil with terrorists armed with chemical weapons. Although military planners have questions about such U.S. air attacks, they have no problem making the hawks in Tehran fearful. There's no downside to letting the hawks contemplate the devastating costs of American reprisals. We should let them—always as part of a larger package that includes diplomacy and economics, unaccompanied by the type of empty public threats voiced by the Bush administration. In sum, military force has several specific tasks it can perform regarding Iran, all potentially useful, but none decisive—that is, none that would compel Tehran to humble itself to U.S. demands.

As for engaging in large-scale combat in Pakistan, that presents a nightmare of inconceivable proportions. No one would recommend it even if the government in Islamabad fell to the worst extremists and gained control of Pakistani nuclear weapons. Washington certainly would warn such a government of the terrible consequences of harming us, but would not start a war. Deterrence would have to be the order of the day.

Pakistan, Iran, and such actual or potential nuclear-armed smaller countries are the toughest cases for the use of American force, whereas the next level down has invariably demonstrated consistent positive results. This level covers peacemaking and peacekeeping operations, usually involving several thousand troops, plus air support for the purposes of preventing genocide and ethnic cleansing, or for broad humanitarian purposes. The enemy here is a government or dominant group within the country bent on ridding itself of people it regards as enemies for reasons of ethnicity, religion, or general "otherness." In Bosnia, the road to military success began after a long period of providing military aid to the victimized groups and backing up this aid with U.S. and NATO air strikes. To keep the peace that was the direct result of this limited but effective use of force, plus the brilliant Dayton Accords, the United States has deployed upwards of 20,000 troops. To this day, they have taken no casualties, and the peace has stood.

To quiet the situation in nearby Kosovo, which was then a province of Serbia populated primarily by Muslim Albanians, required seventy-eight days of bombing of Serbia and Serb targets in Kosovo, plus providing aid to the Albanian insurgents. Again, U.S. and NATO forces achieved their main objectives of stopping the killings and allowing the Kosovo Albanians to move toward independence. In sum, these two Balkan operations can be considered successes primarily because military action cleared the way for peace and keeping all parties to the terms of the peace or cease-fire accords.

United Nations and U.S. troops began well enough in Somalia late in the George H. W. Bush administration and early in the Clinton administration, when their mission was simply to help feed starving Somalis. But later, when Clinton expanded the mission to quelling Somalia's internal tribal war and to a nation-building enterprise, the troubles began. The peacekeepers simply didn't have either the military capability or the will to absorb the casualties or pay the costs of getting this job done. And at the first hint of tragedy—when U.S. troops were ambushed and killed in the Somalian capital of Mogadishu in 1993—Washington quit the operation entirely.

However, U.S. and Latin American peacemakers fared better in Haiti. The approximately 20,000 U.S. forces that were dispatched to Haiti in 1994–1995 achieved our stated goals: removing a dictator; restoring Jean-Bertrand Aristide, the elected president; more or less stifling gang warfare; quelling the general anarchy; and stemming the flow of Haitian refugees to Miami, which had burgeoned into a serious problem.

In Rwanda, the failure and the tragedy were unspeakable. But the problem was not misuse of force, but rather the failure to use force. Washington wouldn't send its own troops to quell the raging tribal genocide there and even stood in the way of others assuming this responsibility. Clinton apparently feared that if others got involved and got into trouble, the United States would be compelled to go to their rescue. After Somalia, he didn't want to send in troops and have them become embroiled in yet another civil war.

These are instances of what has come to be known as humanitarian intervention, with the presumption being that such intervention doesn't really involve vital or even important strategic U.S. interests. The rationale is that U.S. intervention would be based solely on the humanitarian concerns of Americans. I believe that in each and every one of these cases the humanitarian reasons for our intervention were paramount but didn't tell the whole story of American interests.

Standing up to genocide is a critical American interest in and of itself because failure to do so will inevitably lead to the spread of violence elsewhere and to the rise of refugee issues and economic problems. Those spillovers do touch our economic and material interests. In addition, states being torn apart in this way become failed or failing states, which in turn become breeding grounds for terrorists. And finally, and of great importance, American inaction in the face of such consummate evil would lead to the most profound cynicism within American society itself, and nothing would undermine our democracy—the heart of our national security—more than that. The failure of the United States and other nations to act in Rwanda certainly made it much more difficult to deter the genocide in Darfur and elsewhere. To further undermine American credibility and moral standing, the leaders of the Bush administration referred to what was happening in Darfur as "genocide," a term which legally obligated the United States to take action to stop it—and still we did nothing of consequence.

One of the main arguments against U.S. military intervention in these situations is that to stop the ethnic cleansing or genocide, we would have to insert American armed forces over an extended period of time in a civil war or in an extremely difficult, if not impossible, nation-building process. But that clearly has not been, and need not be, the case. Washington can take a variety of military actions far short of sending troops into internal, open-ended combat. In Bosnia, the United States should have provided arms to the Croatian and Muslim victims from the outset. That alone would have helped establish a balance of power on the ground that very well might have led the Serbs to cease and desist their killing and negotiate far earlier than they did. As it was, the Serbs were taking hardly any casualties, because of American and European denial of these arms to the Bosnians and the Croats. If more were needed than that, the United States could have provided air strikes against Serb units, yet another incentive for the Serbs to stop and negotiate. In fact, when the United States and NATO ultimately provided those arms and carried out those air strikes, the Serbs quickly agreed to negotiations and the war soon concluded. It was not a massive commitment of U.S. forces to open-ended conflict, and the follow-up did not require massive economic aid for nation-building.

In Rwanda, again, there was no need to pose the choice as either going in full-bore or staying out entirely. The Pentagon itself offered the idea of sending about 5,000 U.S. troops to border regions in Rwanda to carve out safe havens for those being hunted. This could have saved huge numbers of the victims without significant numbers of U.S. forces ever being dragged into the civil war itself.

I can accept that certain realist hardheads do not share my sense of the importance of American interests in humanitarian intervention. But surely the record demonstrates that these interventions have been important enough to justify the use of limited American forces for limited periods at reasonable costs. These missions have been generally effective, to say nothing of the value they have added to America's standing around the world.

There's a final category of military action that entails eliminating bad and weak dictators with a small and swift military stroke. The two most recent examples have been in Grenada in 1983 and Panama in 1989–1990. Neither of these actions was without difficulty or incident, but both essentially achieved their purposes. In Grenada, U.S. forces quickly and successfully rescued up to 1,000 Americans and drove out the Cuban-leaning thugs who had taken over the government. In Panama, about 26,000 troops overthrew Manuel Noriega, an unpopular dictator in a generally pro-American country, in about two weeks.

To review the record of the use of U.S. forces at various levels and in different types of combat over the last two decades is to see a complex picture. But several conclusions emerge.

First, having overall military superiority has paid off, given the wide range of situations in which presidents have opted to use force to deter anti-American actions; it made sense to err on the side of more military capability rather than less, and certainly to maintain America's technological edge. This record places the burden of proof on those who would cut the defense budget, to show that current dangers would not be far more difficult to face without American military advantages. Do the skeptics really believe, for example, that Iran and North Korea would be more pliable if Washington were weaker militarily? Do they believe that the Taliban would have been driven from power in Afghanistan by the United Nations' tooth fairy or the European Union's devotion to international law? For the foreseeable future, it's fair to say that the question will not be whether the maintenance of U.S. military superiority is wise, but rather how to use that superiority more effectively and in concert with other forms of power.

Second, in those instances in which American leaders called on this military superiority to wage long-term and high-intensity combat, as in Iraq and Afghanistan, military superiority did not translate into stable victories. The U.S. military superiority was enough to dissuade enemies from fighting us directly, battalion to battalion, but not enough to prevail countrywide. Winning or losing countrywide required military superiority and good strategy, but also went well beyond those factors. Conquering capitals is one thing, and conquering countries and keeping them conquered in relative peace is quite another.

Third, the costs and consequences of the wars in Iraq and Afghanistan and of potential future wars involving Iran and Pakistan are great. The outcomes are uncertain. Such wars should be undertaken only when the United States has well-founded confidence that allies within those countries are worth fighting for—that is, that they will be relatively honest, efficient, and at least as skilled and motivated to fight for themselves as our troops will be in fighting for them. When we doubt the caliber and honesty of those we're trying to save, when we believe that they are unable to inspire their own troops to fight with a zeal equal to that of their enemy, our strong presumption should be to avoid combat. Other, more modest strategies—providing aid and training to friends or deterring and punishing enemies—can suffice to meet America's bottom-line needs. Deterrence has worked on almost all occasions when presidents positioned it clearly and firmly.

Fourth, America's present military capability is sufficient to parry dangerous threats and to carry out a range of humanitarian interventions. These involve limited and restricted uses of force and can be fulfilled in quick-hit assignments.

Fifth, military force alone is unable to solve any of the major military threats and challenges now faced by the United States. To manage the toughest situations will demand attention to America's military power, the poised and posed sword, as well as nonmilitary power. Indeed, our military capability and our military prowess can best fulfill their promise in the twenty-first century by being exploited mainly as military power—that is, to deter, to induce negotiations, and to represent credible punishment.

IF, AS WE'VE SEEN, military power in a power package often succeeds where military force alone fails, you have to wonder why American leaders have so readily and so often resorted to war or why they believe they have no choice but to use it. I ask this because of the following considerations: Most adversaries know well that if they seriously threaten us, odds are that we will employ our

arms. It's happened so many times already. Others are also familiar enough with America's formidable military capability to realize
that if we unleash force, they will be likely to lose power. Except
for Vietnam, that's been the outcome time and again—in Baghdad,
Kabul, Panama, Grenada, Haiti, Serbia, and so forth.

Thus, if enemy leaders believed that we had the capability to
clobber them, were likely do so, and would prevail if provoked, they
should have concluded that the story would end with their hiding
in spider holes like Saddam. That denouement, in turn, should have
provided the necessary incentive for them to have capitulated, or at
least to have compromised to our satisfaction. But they hardly ever
did, and we wound up sending in U.S. armed forces. Why?

The first and most obvious explanation was that they didn't believe America would attack, whatever its past record. "They did it to
others, but won't do it to us," they often seemed to reckon. In other
words, in those instances, presidents failed to establish credible jus-
tifications and explanations. They failed to say the words and take
the military steps to convince the enemy that the White House was
serious. Jimmy Carter virtually removed the threat of military force
against Iran as a means to free American hostages. By his supercau-
tious rhetoric, Bill Clinton all but took one of his own options off
the table in Kosovo: the possibility of using ground forces. These
presidents, in effect, gave away the power bang or the scare factor
proffered by America's history of frequent military interventions.

Second, and less obvious, such credibility might have existed initially, but then presidents such as Carter and Clinton lost it. Hai-
tians certainly worried much more about American intervention in
their internal war before the United States scampered out of Somalia
than they did afterward. Remember the Haitian mob that gathered
on the dock in Port-au-Prince, shouting "Mogadishu" (the Somalian capital where U.S. troops were ambushed and had retreated) as
they sneered at U.S. destroyers approaching their shores. Ronald
Reagan, who had taken great pains to stoke his credibility for being
capable of taking tough actions, sullied his war mask when he removed American troops from Lebanon after having called their de-

ployment "vital." And beyond doubt, George W. Bush shredded his credibility with the Iranians, the North Koreans, and others by getting bogged down in two inconclusive and increasingly unpopular wars, and by repeatedly threatening not to tolerate this or that move toward nuclear capabilities—only to do nothing. The surest way to squander credibility is not to do what you threaten to do.

Finally, the least obvious conclusion is that our threatened enemies might well have begun by taking U.S. threats seriously, but saw no reasonable way both to meet American demands *and* to remain in power. In other words, if the adversary believed the choice advanced by Washington was "Get out of town or we'll drive you out," he calculated again and again to risk staying and see if he could absorb the military blows and survive. The best example of this was Saddam's decision to secretly rid himself of WMDs in order to avoid an American attack, in the apparent hope that the Americans would gradually discover that he had done so and would therefore not attack. But he also apparently believed that he could not openly acknowledge his nuclear nakedness and still retain any chance of staying in power. From Saddam's perspective, Bush's approach was all sticks and no carrots, which gave him little choice but to hold on and see what would happen. If Bush had offered a reprieve to Saddam and a gradual lifting of the economic sanctions, Saddam might well have unmasked his empty WMD storehouse and permitted the necessary inspections to verify it. It's possible and even likely that had Bush combined threats of military force with positive diplomatic and economic power, two things would have occurred: Saddam would have publicly confessed the emptiness of his WMD threats; and Bush could have claimed victory without war. The price tag would have been Saddam's keeping power. The benefit would have been Iraq's remaining as a counterweight to Iran.

That's what a power package—military power with economic and diplomatic power—can accomplish. Just such a package was certainly instrumental in persuading Muammar Qaddafi to abjure terrorism and his pursuit of WMDs, and at least for a while persuading Kim

Jong Il to take tentative steps toward denuclearizing as well. In both cases, it was essential to put aside political hyperbole and recognize that neither country would back down simply when confronted with American military power. Threats of force alone don't move dictators. But when the threat of force is combined with an expressed or implied willingness to live with the dictators, even to recognize them diplomatically, and to provide economic benefits or make economic opportunities available, the negotiations begin to make progress. Washington was finally giving these regimes a way of complying with its major demands—namely, that they dismantle their nuclear programs and stop supporting terrorism—without also insisting on their political or bodily demise. In fact, in the Libyan case, the Libyans themselves took the initiative in presenting such a package to the British and the Americans.

So far, the Libyans appear to be living up to their part of the bargain, but it is more difficult to predict North Korea's next moves. Qaddafi may well have been worried about his grip on power, although apparently he had a sufficient hold on it to give him the confidence to negotiate and make significant compromises. The North Koreans, isolated from the world, seemed terrified that showing any weakness and giving foreigners any foothold in their country would eventually destabilize their regime. This probably helps explain the on-again, off-again nature of any agreement with this regime.

As this recent military history shows, Washington often scanted military power and resorted to war because its threats lacked credibility; the United States waved only the military stick and failed to add political and economic carrots and sticks to the power package; or our leaders were artless in employing military power to avoid the dread and costs of war.

AFGHANISTAN IS THE MOST pressing place to examine the possible benefits of the power package, including military power. Washington has nearly reached a consensus that the situation there requires more troops, both American and NATO; more economic aid; and much more political reform. In all likelihood, this will be tried for several years, perhaps as long as a decade, with, at best, an uncertain outcome.

The power package strategy won't receive much of a hearing, even though it offers a chance to meet the most basic U.S. security concerns without greater commitments. Politically, this would entail offering a decentralizing power-sharing arrangement to the Taliban, whereby they would control the southern provinces in return for an end to their harboring of al-Qaeda there. Militarily, it would focus our military power and other instruments on deterring the Taliban from once again becoming a base for terrorist operations and, should they relapse, on punishing them with devastating air strikes. We can also make life difficult for them by providing military aid and intelligence to the warlords who oppose them domestically. Economic carrots can be added to this package as well, over time and as reward for their complying with demands.

There will be those who say that the Taliban can neither be politically cajoled in this manner nor militarily deterred. Maybe so, but I still believe that it's worth exploring. The attempt doesn't look so bad when silhouetted against the present inconclusive and costly course.

The same power package has decent prospects with Iran as well. The mullahs and clerics in power in Tehran have their own historical reasons to guard against American influence. Although Americans today barely remember Eisenhower's overthrowing a legitimate government in Tehran and America's long embrace of the shah, these remain daggers in Iranians' memory. That said, some segments of Iranian society are very concerned about conditions in their country and have reached out to the United States with a comprehensive proposal for negotiations similar to the ones Qaddafi put forward more than a decade ago. This unsigned and unofficial memo, passed to Washington through the Swiss in 2003, offered to meet American security concerns, including some compromise on Iranian uranium enrichment programs, in return for American economic and security help. To this could be added cooperation against the Taliban, as occurred at the beginning of the Afghan war, as well as a role for both countries in the future of Iraq.

In time, this approach can reap rewards with Iran. The Iranian population is undoubtedly more pro-American than any other people in the region and better prepared for democratic government. But whether any Iranian government would give up its right to have a uranium enrichment program is highly dubious.

There are various compromises that can accommodate essential interests here, and once an overall package of carrots and sticks is on the negotiating table, there will be progress. Again, the point is clear: vast U.S. strike capability in the area surely checks Iran to some degree, but alone, it has not solved and will not solve the Iranian nuclear problem or other American concerns. The power package has a reasonable chance of success and is worth trying.

Pakistan is perhaps the biggest security threat. To begin with, it's virtually inconceivable that the United States would be able to defeat and neutralize the Taliban in Afghanistan as long as they have a safe haven in Pakistan. Yet the Pakistanis won't deal with that problem themselves and have openly and strongly criticized American military forays into their territory. The government in Islamabad has been highly reluctant to fully take on the powerful tribes in these northwestern provinces, regardless of U.S. aid to Pakistan and whatever the consequences for America's position in Afghanistan. The Obama administration has stepped up these cross-border operations, with uncertain and volatile effects on relations with Islamabad.

No U.S. president is likely to warn Pakistani leaders to either deal with the Taliban, or let the United States deal with the Taliban, or, failing both, forget about our billion-dollar U.S. economic aid program. Bush did not make such threats for fear that the Pakistani regime, with its nuclear weapons, could fall into extremist hands, and neither has Obama.

This is precisely the kind of dilemma presidents have faced for decades. On the one hand, we absolutely need a friendly regional government to cooperate with us or we run a high risk of failure. On the other hand, if we push that friendly government too hard to help us the way we want, a far less friendly government could take its place.

No military force and no clever power package can provide the answers here; only policy can. The only way to fix a problem where we are damned if we do and damned if we don't is to escape the dilemma itself and get out of the situation. To me, the choice comes down to this: Either go full-bore in Afghanistan and press Pakistan for essential help against the Taliban, or deter the Taliban in Afghanistan and focus on helping Pakistan move toward more political and economic stability. To me, the former is a high risk with a low prospect of success, and the latter is still risky but with a better prospect of success. One point is certain to me: We cannot let the stay-the-course crowd make the United States stay every one of the present courses plus Iran plus pushing back on Russia. All that is way beyond America's military might and power—and its vital interests.

MILITARY POWER, LIKE MILITARY force, is coercive, but it doesn't require physical destruction. Its aim is to change thinking and behavior by putting at risk what adversaries value most. Military capability—our vast arsenal of weaponry and our standing armed forces—contributes to military power. Others see it and imagine what it can do to them. Military force successfully employed in terms of costs and results also adds to military power. It has demonstrated prowess. But military power itself is essentially psychological—compelling others to ponder and worry about how you will act, and to assess the costs and benefits of meeting your demands in order to avoid your use of force. Its purpose is to cause your adversary sleepless nights, but without further stiffening his will or further uniting his country against you, or provoking him into war. As we saw with Saddam, if your adversary thinks your aim is his total capitulation, the chances are he'd rather fight than switch.

Most American leaders have not mastered this art. Nor did they work on the fundamentals that would give military power the chance to flower.

First, the best offense in today's world requires a strong defense. This means Washington must become far more serious than it is today about homeland defenses. To be credible in terms of throwing a punch nowadays, the United States must be able to take a punch from terrorists and even possibly from weapons of mass destruction. For today's adversaries to believe that America genuinely intends to punish them, they also have to believe Americans can quickly recover from the counterblow. Nothing can contribute more to the credibility of our military power.

America's homeland security needs improvement urgently. It took New York City, the Pentagon, and the country as a whole months to return to normal after 9/11. Right now, recovery from a large-scale conventional or WMD attack against major cities, trains, and ports would take much longer, most likely years. Homeland security is now being treated in Congress as the equivalent of pork barrel highway projects.

What's more, even if the chances of future attacks are estimated to be low, no money spent on homeland security programs will be wasted. These are infrastructure programs—ports, hospitals, bridges, communications equipment for police and firefighters—and they will return enormous economic value and societal benefits. Such improvements always more than pay for themselves in their multiplier effects on the overall economy. Of preeminent importance to bear in mind is that the American economy is the basis of American national security power. It is from this that we can afford our military capability. In any event, Americans need and deserve real protection and the capacity to recover from attacks.

Second, the United States needs to establish a credible internal process for reviewing and deciding on war—or run the risk that foreign leaders will convince themselves that the president can't or won't pull the trigger. Some restoration of the formal congressional process of declaring war, for all the controversy that it entails, is probably the best way both to avoid ill-conceived wars and to prove that the country is behind presidential threats and won't desert a president when things go awry. Hawks decry placing limits on the presidential war power, yet they offer nothing to deal with the problem of the waning of will to wage war at home when wars seem endless. If hawks really believe that lack of will at home is the root of all America's military weakness, there's no more reliable means to assert the necessary will than with a congressional declaration of war.

A modernized version of the process would be needed. Congress and the president would also have to devise guidelines to deal with difficult exceptions, such as when the president absolutely must act quickly.

This formal process can enhance military power. It would reduce the chances of badly conceived wars, as would any open, disciplined, and solemn process. It would also be a powerful deterrent, as malefactors would have to gaze upon this gathering of national resolve. It would further reinforce perceptions about America's will to stay the course. Even though Americans usually do, others don't think we will. Finally, the declaration process slows down the rush to war and gives more time for military power and diplomacy to kick in.

A third way for presidents to enhance military power is to stop issuing constant threats they either can't or don't intend to honor. George W. Bush threatened Iran and North Korea endlessly, yet he took no military action against either.

Power also dwindles when weak, pro forma military actions are taken. Clinton slammed a few Baghdad buildings with missiles, and everyone concluded that this was all he intended to do. Vague threats *sound like* vague threats. Sometimes, threats ring hollow without the backing of a particular ally, as when North Korea failed to buy Bush's threats, knowing that South Korea was not on board. And as with most exercises of military power, threats must be framed to allow the target both to comply and to save face.

American leaders could also do well with a dose of creativity regarding military power, i.e., being more imaginative in taking good advantage of the array of military options short of force. Take, for example, China's shooting down one of its own satellites in 2007. Washington didn't like this one bit, but what more effective means of demonstrating that Beijing had the capability to neutralize America's unique war-fighting advantages in information and command and control?

And none can forget Jimmy Carter's missed opportunity to exercise strategic power at the outset of the hostage crisis in 1979. He eschewed the use of force entirely, except for his one strange, lame rescue effort near the middle of the crisis. Instead, he could have given an ultimatum: either release the prisoners or the United States will bomb Iranian oil and port facilities one by one, day after day. Most countries would have stood with Carter because the new Iranian government had committed the unpardonable sin of grabbing hostages from legally untouchable diplomatic soil.

A major opportunity for creative military power received little attention in the second war against Saddam. The idea was to separate southern Iraq from Saddam and his Baghdad power base, in the same manner that Washington had divorced and protected the Kurds in northern Iraq from Saddam a decade earlier. After the First Gulf War, Washington gave air cover to the Kurds in the north, who effectively established their own quasi-autonomous region. In 2003, the same could have been done with the Shiites in the south, who were violently opposed to central Sunni rule, just as the Kurds had been. It would have taken perhaps up to 25,000 U.S. troops plus air power to keep Saddam's troops at bay. In addition and critically, the Kurds and Shiites together would have controlled almost all the oil, the sole source of Iraq's wealth, and thus could have brought Saddam to his knees economically, and perhaps have ignited a coup against him as well. This could have spared Americans and Iraqis a war lasting six years and counting.

All these complicated calculations can be boiled down to a few rules.

RULES ON MILITARY POWER AND MILITARY FORCE

Rule 1: Buy the biggest and most potent military capability you can afford, both for the message this will convey to friends and adversaries alike and for the options it will provide to address the uncertainties you will face.

Your military does not like to make tough choices on what to buy and will always want a lot of everything, particularly the good old ships, tanks, and aircraft. You'll want to have U.S. capability that zeroes in on the kind of situations and wars you deem most vital. You won't be able to buy everything, but don't give in on the need to maintain overall military superiority.

Rule 2: Before you smash anyone with military force, take your time, and then even more time, to be as convinced as possible that the threat warrants force, that there is no alternative path to explore beforehand, and that you can handle the inevitable traps and hurdles at a reasonable cost. Tell those who demand instant action to go to hell.

In addition, be sure the parties you seek to save are worth it: i.e., that their own people will fight and die for their cause. If you don't have confidence that they will, put the blame on them and don't get involved. Take the measure of whether your allies will join you with significant support. If they won't, you might wonder who's being stupid—they or you. And finally, look hard at whether there are reasonable compromises you can make to live with our country's enemies. Keep in mind that we're now cohabiting on decent terms with Vietnam, Libya, China, and Russia, to name only a few.

Rule 3: If you decide to wage war, give our military all the military punch they can justify, but be sure to grill them on their strategy for fighting. Of equal importance, don't let your senior aides leave the building until they've put key political and economic policies in order and coordinated them with those of the military planners.

If you're going to send Americans off to kill and be killed, promise your mother that you will personally toil over the strategy, and not delegate this enterprise—even to Sun Tzu, Machiavelli, Bismarck, or Kissinger. This will be your war.

Rule 4: Military power—tied to economic and diplomatic power—is the best means of getting the most bang out of your military buck without war.

Think about how best to use your superior military capability and demonstrated military prowess for purposes of wielding military power. And never think of military power in isolation; always think of it as locked in a power package that includes economic, diplomatic, and political instruments as well.

These are your most promising instruments of coercion and pressure, and you must move them all into play on vital issues. While the prospect of American forces bearing down on a detested enemy can be truly bracing, never forget that military threats by themselves rarely work. Economic punishments and rewards help as well. And face-saving exit strategies for the enemy's leaders are usually also necessary so that they can both meet your bottom-line demands and retain power.

For better or for worse, that's the complex kind of deal military power and the power package can deliver. It's almost always worth trying this package before making war—unless you're confronting a Hitler.

The ancient Chinese called this power package of threatened punishments and proffered rewards the "golden bridge." Americans—raised on a history of wars of annihilation and unconditional surrenders—shun golden bridges as gateways to hell. But so are wars that don't have to be fought.

CHAPTER 9

ECONOMIC POWER

*You have to think of economic power as the tide and military power
as the storm. Economic power doesn't provide quick fixes; it is rather
a steady, relentless force that eventually works its way and produces
fundamental changes.*

The response of both America and NATO to the Russian invasion of Georgia in 2008 was tepid and fearful.
A tough rejoinder, they reckoned, might trigger further
assertions of Russia's military superiority on its borders and Russia's strength as Europe's main supplier of energy. But Moscow had
an economic jolt coming, one that it did not expect at all, and that
Western leaders themselves had not foreseen: Private investors from
around the world began withdrawing funds from the Russian stock
exchange, leading at one point in September to a reported 40 percent decline in value from its highs earlier in the year, and a sharp
drop in the value of the ruble. Investors in Russia clearly had second
and third thoughts—not necessarily because of any philosophical
opposition to Russian action in Georgia, but because they now saw
Russia as more capriciously governed and a riskier investment.

The market, not governments, had imposed its own sanction on
the recently re-empowered Russia. Private investors had compelled
Kremlin leaders to think about the adverse economic consequences

of further aggression. And for all its new oil and gas wealth, Russia badly needed foreign investment for modernization. But now, the Kremlin had created uncertainty about whether the government was dependable; the Kremlin had made Russia a poor risk for credit and investment. Perhaps, private investors, acting in their own economic self-interests, and not controlled by their governments, had inadvertently found the power key that had eluded their own political leaders—a way to deter future Russian aggression, or at least, to make Moscow recalculate the costs of exercising its new power.

This stunning and surprising event calls to mind the new mantra of international specialists in the twenty-first century: that economic power now occupies center stage in world affairs. The mantra is accurate. It is even chanted in the United States, where it is, alas, far more preached than practiced. The linking of trade, investments, and resources to foreign policy and military affairs has been second nature to most nations for centuries. But this has not been the case in America, where principle and politics unite to "protect" economics and business from government intrusion (except when needed), where the departments of State and Treasury still avoid collaborating on policy, and where intellectual apartheid separates economics and politics departments at universities.

Two contrasting experiences in my own life have made me particularly aware of this unfortunate division in American thinking. The first was a positive experience when I was a graduate student at Harvard, observing many of our nation's best minds actually creating the new field of national security and arms control in the late 1950s. The second, a negative experience, began in the 1990s, when I assumed the presidency of the Council on Foreign Relations and tried, unsuccessfully, to develop a program combining foreign policy and economics.

It was evident in the early days of the Cold War and the nuclear era that traditional thinking on foreign policy couldn't accommodate many of the new international challenges of that era. To address them, groups formed at the RAND Corporation in Santa Monica and among professors at Harvard and MIT in Cambridge. They included foreign policy specialists such as Henry Kissinger and Albert Wohlstetter, economists of the order of Thomas Schelling, chemists with an interest in international affairs such as Paul Doty, and physicists like Herman Kahn, as well as law professors, lawyers, historians, and others. They held seminars, published books and articles, founded new journals, worked with graduate students like me, testified before Congress, and urged the creation of the Arms Control and Disarmament Agency and related bureaus throughout the government. They served in government, returning to private life to publish even more. From this stunning, multidiscipline intellectual festival emerged the new and critically important field of national security and arms control.

I tried to replicate this intellectually stimulating experience when I was president of the Council on Foreign Relations. Just as most specialists in the 1950s had recognized the need for the new discipline of arms control and national security to combine military and foreign policy, so they now—in theory at least—called for a blending of the study of foreign and economic policies. I raised the millions of dollars necessary to fund such a program, mainly from the New York business and financial communities. I contacted the relevant departments and graduate schools at Columbia and Harvard Universities, and pushed and pushed—with little response. I was able to give away only two scholarships for future economists to study foreign policy and for future foreign policy experts to study economics. Academic walls, inspired by professorial theorists who rule their fiefdoms, had risen far higher than they were in my graduate school days. All the while, the Clinton and Bush administrations would not make great headway in meshing the gears of government in these two fields.

We still have no academic field comparable to the political economy courses in Europe. And, as a country, we still don't have much of a foreign economic policy to exploit our formidable economic power. One of America's best potential sources of international influence is underutilized, and given the subtleties of how economic power works, this problem won't be easy to fix.

WHILE A REVELATION TO some American foreign policy experts, international economics is not a recent invention. Trade and wealth have always both inspired the flag and followed it. Precious metals, exchanges of goods, bribes, and the quest for riches always preoccupied tribes and nations. But it is different today. For the first time in history, economics has become the principal coin of the realm, displacing military and diplomatic power. Few would dispute its ascendancy, but it happened relatively quickly in historical terms, and the new role of economics and its banquet of meanings and uses are still being digested.

China, so recently considered primitive and backward, is now among America's major trading partners, despite the two countries' underlying rivalry. As of December 2008, China reportedly sat on $1.9 trillion in foreign exchange reserves. Now, Russia, only two decades after utter military and economic collapse, is, once again, a major diplomatic player by virtue of oil and gas exports. Western European states have drastically downsized their armed forces and now keep their hands on world tillers mainly by employing the market attractions of the European Union (EU). In 2008, daily financial transactions amounted to more than $3 trillion—compared with billions in past decades—mostly beyond the intrusive hands and understanding of governments. At the outset of this economic blossoming, President Clinton gazed upon it, pronounced it a new era, and believed it promised the end of all wars and world poverty and thus the advance of democracy.

Only now are American leaders beginning to see that most economic instruments don't lend themselves to the traditional day-to-day exercises of power—to squeezing, pressuring, or coercing others. Trade, aid, investments, bribes, and sanctions work more slowly than does military power. Indeed, these economic measures and activities usually don't have any short-term political impact at all, except possibly in the case of bribes. The promised rewards and punishments take time to seep in and grip the lives of others.

Also of critical importance, governments generally have far less control over the wide range of economic instruments than they do over military power. The economic activities that carry the greatest weight, such as trade and investment, rest more in private and corporate hands than in the clutches of governments. By contrast, the state has virtually total control of military force.

For these reasons, it is better to think of economic power as the tide, a force of gravity that inexorably pulls things to and fro and reshapes shorelines in the process. It is thus better suited for long-term strategic positioning than for quick tactical gains. The creation of wealth is an activity in which all countries want to participate. In the era of globalization, it is one of America's major attractions. American leaders often mistakenly try to treat economic power as if it were military power—as just another instrument of pressure. This demonstrates that the United States—unlike China and many other states—still lacks a good handle on economic power.

ECONOMICS HAS NOT BEEN a part of the comfort zone of traditional power masters. To be sure, sages like Thucydides understood well the power of gold, especially to sustain armies. But you need a microscope to spy other than passing references to economic matters in writings by legendary figures of foreign policy such as Machiavelli, Otto von Bismarck, and Henry Kissinger. They appreciated that great states required wealth and economic growth. But to them, real statesmen didn't do economics. To them and to the kings who forged empires, riches were essentially a means to military might, and military might was the road to international power and security—and as a bonus, to even more gold. To them, foreign policy was strategy, and strategy swirled around military power and diplomacy.

Today economic power is the principal path to international power, the entry ticket to the international power elite. Military storms in the twenty-first century are generally deemed to be excessively costly and riskier than ever, given international norms, domestic needs, and uncertain results. Economic power goes to the heart of what now counts most to most countries—the economic welfare of their governments and their citizens—and it's infinitely less expensive than war. And while economic power works more slowly than military power, the changes wrought by economics are surer and more lasting. If used strategically and with patience, it could be the most cost-effective means of persuading other governments to do what they don't want to do—and of persuading them without war.

ECONOMIC HISTORY DID NOT really begin to reach the West until the Renaissance. Before the Renaissance and before the later industrial revolution, the Chinese and Indian economies were much stronger than those in Europe. China and India led in science and new technologies like movable sails and gunpowder, and in trading skills. Europe resembled a horde of barbaric tribes. Only after 1500 did Europe transform itself into a garden of nation-states and really launch the new era of nation-states and empires. More so than Asians, the Europeans had organizational skills and practical know-how, which they converted into superior military might and, in time, economic growth.

Superior wealth was indeed a prize and, indeed, to many, *the* prize. This was recognized especially by the mercantilists of sixteenth-, seventeenth-, and eighteenth-century Europe, who glorified the pursuit of wealth through control of foreign resources and trade surpluses. Arms would control resources and impose rules of trade, which would create wealth, which would lead to even greater military might, and so on. Mercantilists were, in effect, economic nationalists. The Ottoman, Spanish, Portuguese, and British empires, among others, followed similar patterns.

From roughly the 1800s until the guns of August 1914 came a period that has been called the "first era of globalization"—a period during which the major powers of Europe reached new heights of mutual trade and investment within Europe and beyond, attaining proportions marveled at even today. Perhaps some sensed the forces of conflict lurking just beneath the surface, but this did not prevent a temporary quieting of national rivalries and a rush toward mutual dependence—for that brief period.

The emerging economic interdependence seemed so great as to inspire some to predict the end of war itself. War, they argued, had become too expensive, unaffordable. Socialist parties and labor unions even went so far as to publicly forswear raising arms against their fellow laborers in other countries. Of course, these dreams of wealth and perpetual peace faded into the cannon roars of 1914.

The war's end returned the spotlight to economics, but in a punitive way. The victors imposed harsh reparations on Germany. They also tried to weaken the new communist Soviet Union by restricting trade. In 1939, Washington resorted to this instrument—cutbacks in vital raw materials—to show its displeasure at Tokyo's invasion of China. The reparations only embittered the Germans, and the sanctions against Tokyo only infuriated the Japanese.

America's wondrous manufacturing performance during World War II, and the fact that the United States was the only victor standing with a strong economy at the war's end, gave Washington enormous power. The Truman administration used this standing to fashion a foreign policy that, for the first and only time in American history, featured economics. With this policy, the United States would lead the world to establish a series of new multilateral institutions such as the World Bank, the General Agreement on Tariffs and Trade, the United Nations, and of course the Marshall Plan. Truman rejected the policy of reconstituting the American armed forces he had so recently dismantled and also rejected making use of his temporary monopoly of atomic weapons.

Truman's international economic institutions remain today, but by the 1960s, military power had overwhelmed national security policy. The blessings of commerce and its profits were subordinated to security concerns. The military budget started to climb. Washington even granted generous aid to countries for the privilege of defending them against communism. For good measure, Washington tried to squeeze the Soviets and the Chinese by restricting the world's economic transactions with them. For its part, Moscow attempted to placate its poor and unhappy Eastern European satellites by subsidizing oil sales to them.

Decades later, economics reentered the big strategic picture under President Ronald Reagan. He and his supporters made their own conceptual breakthrough in economic and security policy, which they insisted was instrumental in winning the Cold War. Their plan was to ratchet up military spending, especially on a high-tech, high-cost missile defense known as Star Wars, and drive the Soviets into economic bankruptcy. The theory was that despite their flagging economy, the Soviets would feel compelled to match U.S. expenditures. In fact, Moscow held steady on its military spending and didn't bankrupt itself seeking to match America's Star Wars efforts. It couldn't and it didn't. At most, the U.S. spending push served mainly to increase the Soviets' sense that they were hopelessly falling behind America.

With the Soviet Union stagnating, then crumbling, and the United States moving even farther ahead, fears of war began to subside and business interests began to gain over security concerns. Washington paid greater attention to competitive economic needs, and to the specific need for the government to help American business as other governments did. In 1989, a major policy shift was signaled in a cable from Deputy Secretary of State Lawrence Eagleburger to all U.S. embassies. It announced that henceforth these embassies would bear a primary responsibility "to develop programs to promote trade, solicit views of resident American business and other private sector people on trade policies and problems, and help meet the challenge of foreign competition. . . . America's economic health is the country's number one national security interest."

With the end of the Cold War, with Russia's fall and America's ascendance as the world's only superpower, came the second big bang in international relations—the era of globalization. Like the first big bang prior to World War I, it was attended by new outsize dreams of economic power replacing arms as the principal currency of international affairs.

Down came the majority of communist and free-world barriers to the movement of money, goods, and technology, all expedited by revolutions in transportation and communications. Up soared demands for more spending within states on welfare and health, and this demand for spending reduced calls for more military weaponry. No state seemed to be threatening another's vital interests, and no major-power wars loomed on the horizon. Worried far less about security, leaders were freed to refocus their resources on economic rather than military priorities, and to spend much more of their time on their economies. All this, in turn, fed into an expanding international economy, already gathering momentum.

In most of the world, Adam Smith and David Ricardo had supplanted Machiavelli and Clausewitz.

IN KEY STATES THROUGHOUT the globe, economic interests came to define security, and economic power became the heart of foreign policy.

Not surprisingly, Russia's overriding goal was to reestablish its power on and near its borders. With current Russian armed forces a mere fraction of what they had been in their Soviet glory days, oil and gas exports emerged as Russia's new strategic weapon of choice. Russian leaders are now in a position to threaten oil and gas cutoffs or price increases, and to invest the country's enormous oil and gas revenues (which account for about 60 percent of its annual exports, or about $390 billion in 2007) in Europe and elsewhere, although their revenues of course declined in the middle of 2008 with falling energy prices. Still, the size of its economy jumped from about $996 billion in 1999 to about $2.2 trillion in 2008 (in terms of GDP PPP).

To convert these profits into international power, Russia's rulers first had to regain government control over their country's previously privately held oil and gas companies. They did so with predictable ruthlessness. Thus Moscow's rulers reclaimed internal control over these resources in order to utilize economic rewards and punishments abroad, much as they once employed the threat of use of force. They moved swiftly to make their points with neighbors such as Ukraine and Georgia, nations that Moscow felt had moved too close to Washington.

Russia has another energy stranglehold on its neighbors. By reducing pipeline access through their territories, it can cut the attendant pipeline fees, which are substantial. And if a neighbor proves particularly nettlesome, the Russians are not above stirring up pro-Moscow independence movements within the neighbor's borders, as they did in Georgia in 2008. They have also moved to blunt European counterpressures by buying up downstream energy businesses, from refineries to delivery trucks, throughout the continent. Europeans resist, but Russia's money is good and its acquisition and ownership continue to expand westward. The bottom line is unmistakable: Moscow has bought itself a new and influential seat at most international negotiating tables and the clout to parry pressure from Washington and Europe.

Saudi Arabia, like Russia, also bases its power on oil, and all the Saudi rulers and their minions have but one goal: the preservation of the rule and power of the House of Saud. They buy their survival with their oil profits, and critically, with the billions of barrels of oil still under their sands. They think that given the West's dependence on their oil, Washington and its allies cannot afford to let the desert kingdom fall into hostile hands. And they're right.

Saudi foreign policy operates like an insurance policy. The regime uses the Saudis' complete control over the nation's oil business to buy off threats. If they feel the Shiites surging in Iraq, they provide money and arms to their fellow Sunnis there. If they see Iran swaying Palestinian parties, a traditional Saudi sphere of influence, they call Palestinian leaders to Riyadh and pay or promise the necessary sums to quiet them down and reestablish Saudi leverage. Interestingly, they actually promise more than they deliver, as can be seen in their very limited investments in Arab or other developing countries. They establish their power among these lesser countries and groups mainly by allowing them to hope for future Saudi largesse. When they actually invest their funds, it tends to be in stable and proven industrialized economies, especially in the United States.

This, in turn, gives Riyadh a double hold on Washington—as oil supplier and influencer of world oil prices and as a major foreign investor in the American economy. But the high-water mark of Saudi influence in Washington was probably reached in the 1980s, with the war against the Soviets in Afghanistan. The Saudis stepped up their oil production to drive down oil revenues to hurt Moscow, and they supplied money and arms directly to the Afghan rebels. Riyadh wants to appear to be the defender of the Muslim faith, but not get out front on anything.

No one has played the economic power game more adeptly than China, and all in the service of perpetuating the rule of the Communist Party leadership. To that end, Chinese leaders faced the strategic choice of buttressing their legitimacy by granting political freedoms or increasing the strength of their economy. They chose seeking greater economic benefits for ever-greater numbers of Chinese. And since China's internal market, unlike that of the United States, was too weak to sustain such economic growth, China chose exports and attracting foreign investments as the main means of creating jobs and of generating wealth at home. In time, its domestic market will expand, but for the time being, its economic strategy depends heavily on world trade, including good ties and robust trade with the United States, though the latter was obviously shaken by the crisis of 2008.

China's international economic policy has both range and sophistication. First, China enhances its economic power by selling its huge economic potential, its future more than its present, its dream more than its reality. This is a shrewd insight into business decisions, and it has made doing business with China the hottest ticket in town, despite significant drawbacks: problematic commercial laws and courts, a weak banking sector, rampant corruption, and environmental disasters. News stories and business journalism often downplay these conditions and fail to mention that China's economy and standard of living remain well below America's. China's gross domestic product (GDP) as derived from purchasing power parity (PPP) is about half of America's ($7.8 trillion versus $14.29 trillion in 2008), and at present growth rates it will remain behind the United States for at least another decade if not much longer. More tellingly, per capita income in China stood at $6,000 in 2008 compared with $47,000 for Americans. Also, the United States still greatly exceeds China in technological innovation and productivity.

Second, China's leaders employ their economic power unthreat-eningly and carefully. Unlike the Russians, they rarely flaunt it; they let it speak for itself. Beijing's diplomats seldom demand explicit foreign policy quids in return for quos like aid, investment, and trade; they just try to drive a good economic bargain—economic value for economic value. China's beneficiaries don't have to be told what's expected of them. Thus most of them shun relations with Taiwan, and rarely condemn China's violations of human rights. Chinese leaders understand the effects of money.

With a variety of economic instruments, China has constructed de facto allies, principally in Africa and Latin America, and including rogue regimes (many of which are oil producers), a group of friends that can be called upon as future strategic needs require. Beijing has also forged multilateral economic organizations in Central and Southeast Asia that exclude the United States and in which China is the top dog. No other nation has matched China's breadth and depth of political-economic activity, or its unobtrusive building of strategic power.

Third and most remarkably, China has made the United States the fulcrum of its international economic policy, even as Washington remains its principal strategic rival. Beijing has, in effect, bet much of its own economic future on America, and to a lesser extent, the reverse is true as well. China's biggest country customer is the United States (in 2008 the United States imported $356 billion from China, though the EU imported $357 billion) and the United States runs its largest trade deficits ever with China ($268 billion in 2008, with the cumulative differential at over $1 trillion since 1995).

Beijing has invested the bulk of its profits in U.S. securities and is eager to buy into private businesses, if and as Washington permits. China is now heavily dependent on U.S. financial markets and on the health of the dollar. The United States' dependency on China for investments and cheap consumer products is also substantial, but is the lesser dependency of the two.

On the surface, at least, it appears that Beijing has locked itself into an inferior economic position with Washington. If relations were to sour, America could adjust more readily because its economy is bigger and more agile than China's. But Chinese inferiority is more apparent than real. Both sides stand to lose significantly (market and jobs at home for the Chinese, and cheap goods and investments for the United States) from a confrontation. Thus neither seeks to test the economic power balance. Both also hold economic plums back for future tussles: the United States has established barriers against Chinese ownership of American businesses, although it did allow the Chinese company Lenovo to purchase IBM's personal computer unit; and China keeps promising to further open its markets and service sectors, and to crack down on its rampant copyright theft (although it has been delivering only incrementally).

Washington is patient with Chinese foot-dragging on many counts, and to some degree, China repays the forbearance by being useful to Washington on tough foreign policy questions such as dealing with the North Korean nuclear weapons program. This leaves both nations locked in a complicated strategic and economic embrace in which, while the present power balance favors the United States, neither can significantly push the other around.

Japan is as intent as China on strengthening its economy through trade, investment, and aid, but its aim is far more economic than strategic compared with China's. In a sense, for Tokyo a thriving economy is an end in itself, while for Beijing it's the means by which the communists can retain domestic power and protect and advance Chinese interests worldwide. The business of Tokyo is business. It's not that business controls the Japanese government, but that the ethos of its government since World War II has been pro-business, and specifically pro-exports and pro-investment. This was reinforced by the anti-defense attitudes that took root in Japanese postwar politics. Japan's foreign policy has been at heart an economic policy.

But its foreign policy has had one golden rule: stay close to Washington with financial and political support, and let the United States take the lead on Japan's security problems, first counterbalancing the Soviet Union, and now performing the same role vis-à-vis China. As one example, to keep Washington happy, Japan broke major taboos in dispatching peacekeepers to Cambodia, and by contributing $13 billion to defray U.S. costs for fighting the First Gulf War. Most recently, Tokyo has been expanding its military capability to hedge against both Chinese and North Korean assertiveness and because of its uncertainty about America's future role in Asia. Its foreign policy status lags behind its economic status as the world's third-largest economy.

European nations and the European Union fit somewhere in between Japan and the United States in their approach to international economic policy. While relations between the government and private industry are not as close in Europe as in Japan, they are generally much closer than in the United States. And while Europeans make more of a pretense of having a military capability than does Japan, the military strength of all twenty-seven nations of the EU is negligible. Europeans maintain just enough military force to claim a place at security tables, but not nearly enough to act independently, except in small missions to Africa. Nations such as the United Kingdom, France, and Germany, along with the EU, exercise their power abroad mainly through their international economic policy and multilateral diplomacy.

Europeans understood the importance of their combined size and took three major steps to maximize their collective strength. First, they expanded the union from fifteen countries to twenty-seven, with a total GDP of $14.82 trillion in 2008, making the EU the single largest economy in the world. While all twenty-seven countries do not always march in lockstep, they do try to present a united front to the world. Second, they established the euro as the common currency for about half of the members. The euro now rivals the dollar as a world currency. And third, Europeans speak with one voice on external trade negotiations—the voice of the EU trade commissioner—and they have achieved genuine influence as a result.

Europeans have enjoyed modest success in translating their market behemoth into significant international diplomatic power by advocating multilateral approaches to global issues and by championing the approach known as conditionality. Conditionality refers to a linking of aid, trade, and membership to the EU itself to political ends—advancements on democracy and human rights. They also use the allure of EU membership to keep the peace between Greece and Turkey over Cyprus, rein in the Turkish drift toward Islamic extremism, and entice Serbian cooperation on Kosovo and other issues.

Multilateralism is Europe's mantra, and it is a highly popular one worldwide. It represents the only hope for most nations to participate in major international decisions. And the EU has somehow maneuvered itself into being the champion of multilateralism. Thus the idea of multilateralism has emerged as the EU's best power multiplier and gives it more of a voice in places like the Middle East.

The EU, like other major economic powers—Russia, Saudi Arabia, Japan, and China—places economics at the center of its foreign strategy. Economics is its main source of leverage in the world. Also like them, the EU uses economic weapons mainly to advance its economic interests and to strengthen its overall regional or international positions. In fact, most major states generally wield their economic power more for economic and strategic ends than to gain tactical concessions on political-military issues.

WHILE THESE MAJOR COUNTRIES have mature foreign economic policies, the United States lags behind, and in fact it is difficult to unearth U.S. thinking on foreign economic policy. When I asked one of the most senior officials in charge of U.S. economic policy since the end of the Cold War how his administration thought about international economic power, he got up from his seat, scratched his head, and responded: "You know, that's an interesting question."

In the new era of globalization, Clinton put economic policy at the forefront of his foreign policy—although the contours of his foreign policy were never clear. President George W. Bush was as strong a free-trader as was Clinton, but his focus and interests, influenced by the 9/11 attacks, reverted almost entirely to security issues and American military power.

The Bush and Clinton administrations both concentrated on U.S. trade and investment as the hallmarks of the new economic era. They started with the correct assumption that the principal lever of U.S. economic power was its ability to open or close the American domestic market to others. Entry into that market was the world's prize in terms of sales and profits. The thinking behind this was simple and well founded in history: In return for being allowed to compete in the U.S. market, other states would open their own markets to international trade, not necessarily as rapidly or as comprehensively, but sufficiently to give American business a fair chance to export to those countries. And whereas this might result in some short-term setbacks, American leaders had the confidence that U.S. business would outproduce, outsell, and outinvest its competitors in the medium and long term. And for the most part, this reasoning held up—until recently. Against new, low-cost competitors such as India and China, the United States started to lose jobs, though its productivity growth stayed solid. Although economists argue over what this portends for the future, the U.S. trade deficit surpassed $700 billion in 2007. The formerly peerless warriors of American business were beginning to fail, and the American economy seemed to some to be less of a prize.

Furthermore, both the Bush and the Clinton administrations were plagued by the fact that the two wings of the U.S. government that managed these affairs—the Treasury Department and the State Department—fell into the pattern of having little to do with each other.

This separation generated a profound clash of cultures—an

almost exclusive focus on finance and monetary issues for economists at the Treasury Department, and an equally exclusive devotion to diplomacy among foreign policy experts at the State Department. Each group of specialists essentially went its own way. And
when the groups had to come together, they rarely shared the same
concerns. The economists at Treasury focused on how much money
to demand while the diplomats at State were concerned about maintaining good state-to-state relations. In other industrial democracies working arrangements were much closer between economic and
foreign ministries, and they could help each other think through
how best to maximize their countries' interests and power.

With Treasury and State, it was mostly a tug-of-war. Under Clinton, Treasury dug in its heels against going after the Serb president
Milosevic's foreign bank accounts. Secretary of the Treasury Robert
Rubin did not believe—as a practical and philosophical matter—that
the United States should be invading the global banking system; he
feared that the sanctity and security of the system would be undermined. Secretary of State Madeleine Albright had no trouble want-
ing to punish Milosevic or any other opponent with this measure.
Similarly, in the George W. Bush administration, Treasury, which
was responsible for combating the international financing of terrorists, moved entirely on its own and without even informing State,
when it seized North Korean bank accounts in Macao. Whatever
Pyongyang was doing with these accounts, its leaders valued them to
the point of boycotting critical nuclear talks with the United States.
Bush's State Department later pleaded directly with Secretary of the
Treasury Henry Paulson, who released the funds. Also under Bush,
State and Treasury locked horns over a new aid program. State
argued in favor of focusing economic aid on failed states as a way
of fighting terrorism there. Treasury countered that giving aid to
states other than those with policy competence, i.e., those that could
make good use of it, was a waste of money. Treasury prevailed.

For all the difficulties of meshing policy within the U.S. government, the economy hummed along during most of the Clinton years
and the early Bush years. Among the bright spots were an increase

in U.S. exports aided by the weaker dollar, as foreigners continued to invest in American securities and firms, and economic growth churned along at low but still positive levels. But the bad news was troubling, highlighted in 2008 by crisis-level defaults in the home mortgage industry and the failure or weakening of key banks and financial institutions, attended by a tightening of capital for investments and a squeeze on credit and consumer spending. All this came in addition to the continuing trade deficit, which surged to ever higher levels. America fell deeper and deeper into debt, becoming the nation with the largest absolute debt. No major power in history had accumulated debts anywhere near this magnitude and managed to remain a major power. Debtor nations must put their fates in the hands of outside investors, who may withdraw their funds as they wish. With this staggering debt came a decline in the relative value of the U.S. dollar, though the U.S. dollar was temporarily strengthened by worried foreign investors during the earlier stages of the 2008 world financial crisis. The prestige and economic advantages of other nations' using the American dollar to denominate their currency values was in jeopardy. Further, individual, private debt was piling up along with bankruptcies.

There was even more bad news from Washington. The federal budget deficit was once again, as in the Reagan years, spiraling out of control. The future of that budget was further imperiled by the impending health and social security costs of the generation born in the postwar years. After years of warnings, dependence on foreign oil was increasing. And most ominously, human capital, America's historic competitive edge, was foundering in a public education system that was scoring ever more poorly on math and science compared with other industrialized and even nonindustrialized nations.

In a very real sense, the United States was losing that which had made it both a great military and diplomatic power and a true world leader: its indefatigable economy, always strong, always a model of economic vitality. This glorious economy was now at risk. The 2008 world financial crisis heightened that risk. Though other leading economies were also damaged by this crisis, and though America still remained the leading economy in the world overall, its absolute economic strength had been sharply diminished. For President Obama—far more than his predecessors—it would become much more difficult to make tough foreign policy choices and to assemble carrots and sticks for the exercise of global power, economic and otherwise.

THE CLASSICAL MASTERS OF power and warfare would be startled at the shifting importance of military and economic power. They would also be surprised to see the decline in the historic patterns of wars between states. In the last two decades, force has been employed principally to quell internal problems or to perform peacekeeping functions within states. The United States, Israel, and Iraq (against Iran and Kuwait) have been the notable exceptions. Many present-day armies are sized well below historic levels, especially in Europe, and can operate only within or near their own borders. But the keen eyes of Machiavelli and Clausewitz would be quick to perceive that finance and commerce now occupy a special and unprecedented perch in foreign policy.

Economic power—whether to reward, punish, threaten, or induce—is like military threats in that it involves pressure, but the similarities end there. Economic power differs from military power in five basic ways.

First, even between economic competitors, there is often a significant mutuality of economic interests, which traditionally was not the case between military rivals. In military confrontations, one side wants to intimidate or beat the other into submission. Wars are zero-sum games. The only shared concern might be to avoid the use of catastrophic weapons that could harm the victor as well as the vanquished. In economic struggles, each side wants to gain, and each can, because winning is generally defined as getting a good or better part of a mutual success. In most economic transactions, there's something for both sides.

Second, the mutuality in most economic relationships makes

it much more difficult, if not impossible, for either side to dictate terms—again, this situation is unlike military affairs. Economic power is much more about bargaining than barking orders. An economic advantage can produce more beneficial results for the stronger side, but rarely a totally one-sided result. In war, a relatively small superiority might be sufficient to achieve outright victory. In trade negotiations, by contrast, even the smallest states can bargain successfully. Both sides have to give and take in negotiations of economic power.

Third, it's easier for governments to control military power than economic power. The fact is that governments in most states completely control their armed forces and are in sole charge of making decisions on the use of force, while they don't have nearly the same degree of control over economic transactions. Whether or how to fight is up to officials of state. By contrast, the power to make most economic decisions is shared among government, private business, and individuals. In most countries, private citizens make most of the decisions on whether to trade and invest, with government officials playing a lesser role—and sometimes none at all.

Fourth, economic power takes longer to work than military power. Military power can destroy a regime with relative speed, but changing the course of its foreign or domestic policies through economic incentives or sanctions can take months or years. The results of economic give-and-take can be uncertain for many years. For example, comparative gains from trade agreements—such as the North American Free Trade Agreement (NAFTA) among the United States, Canada, and Mexico—take longer to become evident.

Fifth, if it is difficult to judge economic pluses and minuses in a strictly economic transaction, it's far more so to evaluate the trade-offs between economic concessions on the one hand, and security and political concessions on the other. For example, it is not obvious whether Bush made a good deal or a bad one in agreeing to exclude India's military-run nuclear plants from international inspection in return for New Delhi's saying it might buy American nuclear reactors. How does one measure the relative value of giving China special trading status in return for Beijing's pledges of improving its performance on human rights? No one has really mastered the puzzle of trading these economic apples for those strategic and political oranges, or, for that matter, of calibrating how much to expect from economic power in general.

NOTHING EQUALS A STRONG economy for economic and other forms of power. It is the magnet for, and principal component of, economic power and the basis of military capability. It is the closest power ever comes to being attractive rather than coercive. A strong economy doesn't fear concessions, which can benefit diplomacy as well. It also allows leaders to wait patiently to collect their winnings, again adding weight to the country's diplomatic powers.

However, when leaders lack patience and lack a vigorous economy, they turn to an array of lesser economic instruments. These come in many varieties: military aid, development aid, bribes, sanctions, trade negotiations, and investments. There is a lot to choose from, more instruments than can be found in any other form of power.

Military aid is a standard policy instrument. In fiscal year 2006, for instance, Washington subsidized arms sales and paid for training programs and mutual visits with the armed forces of about 150 countries. Policy experts often advocate such programs: They bind the U.S. government, or at least the U.S. military, directly to one of the most powerful groups in all small and mid-level countries. In most such nations, the military ultimately decides who holds power internally and has a major influence on foreign policy. Congress sometimes balks at funding these programs, for fear that they imply a U.S. commitment to defend the recipients.

Overall, programs funded with military aid have been positive, sometimes even a big plus. Yet the track record is mixed. Close ties to the South Vietnamese military did not work out as desired in Vietnam, and it is still unclear what the outcome will be in Iraq.

Majorities of citizens in Latin American countries disliked Washington because of its ties with the continent's right-wing military leaders. Military aid to African countries has generally been a mess. In Turkey, the military has been of mixed value, supporting secular governments and being generally pro-American, but with a spotty record on backing democratically elected governments.

The power and good gained by economic aid are difficult to measure. If the aid was multilateral and funneled through international financial institutions like the World Bank or agencies of the UN, it has been particularly difficult to discern exactly what advantage accrued to the United States. But all recipients knew that Washington made substantial contributions and could have vetoed such aid, and it may be that this fact at least had some effect on the recipient countries' attitudes toward the United States. As foreign diplomats and leaders around the world attest, these recipient countries were careful not to offend the United States and thus diminish their chances of receiving aid. And in many cases, if they lost out on multilateral aid, they also became less risk-worthy for private investments. Everything considered, there is some power in economic aid, if and when Washington actually gathers itself together.

Bilateral economic aid, exclusively from the United States to other individual countries, is widely misunderstood. Much of this aid given by the United States is actually paid to American firms (such as American rice farmers) to export their goods and to American consultants, chosen both for their expertise and for their political loyalty. The rationale for almost all the U.S. dollar aid that actually reaches its foreign destinations has been principally noneconomic ever since the Marshall Plan. The major stated goals remain "transformational," meaning that the intention is to modify policies and institutions through the promotion of democracy and free markets; to "strengthen fragile states" (like Pakistan); to "support U.S. strategic interests" in countries such as Iraq and Afghanistan; and to give humanitarian aid and to "mitigate global and international ills" such as HIV/AIDS.

Bilateral economic aid can certainly be credited for the success of both the Marshall Plan in Europe, where almost all of the aid was actually given to the Europeans, and, temporarily at least, the Alliance for Progress in Latin America. Some critics would cite the unhappy days for human rights and democracy in Brazil and Peru in the 1960s and Chile and Bolivia in the 1970s at the very times when U.S. aid to those countries peaked. And economic aid can't be given many gold stars for positive change in countries such as South Korea and Taiwan, where other factors—peace, stability, strong government, private business competence, and trade—earned most of the kudos.

None of this is to belittle the power of bilateral aid, especially humanitarian aid, which engenders positive reactions toward America. The United States virtually always ranks as the nation providing the largest amount of bilateral economic aid, known as Official Development Assistance (ODA). In 2008, U.S. ODA reached $26 billion (Iraq and Afghanistan received the lion's share), with Germany second at $14 billion, and the United Kingdom third at $11.4 billion. And the jury is still out on the success of the Bush-inspired Millennium Challenge Account, whereby aid is given to those nations best able to use it. In general, policymakers would say that this aid makes it more likely than not that a recipient will publicly defend U.S. policies in controversial situations.

When bilateral economic aid, in fact, amounts to a bribe or an inducement for certain behavior, outcomes have been somewhat more favorable. Perhaps the most successful example was President Nixon's offer of billions of dollars to Israel and Egypt for the Sinai accord, which brought about the withdrawal of Israeli troops from that region. President Carter cemented this accomplishment in the 1979 peace treaty between these two countries, pledging approximately $3 billion to Israel and $2 billion to Egypt annually, sums that continue to this day as "budget support." Without doubt, promises of aid in 1994 from the United States, South Korea, and Japan to provide North Korea with light-water power reactors, fuel, and general economic aid pushed Pyongyang toward reasonable compromises. All in all, the record of both aid and bribes is a good one, especially when specific conditions and clear-cut timetables accompanied the bribes.

Economic sanctions almost always cause pain, but don't often succeed in compelling adversaries to yield. Washington has imposed sanctions over 100 times in the last 100 years, including many buttressed by the UN, the majority of these under U.S. pressure. These sanctions include generalized prohibitions on trade, investment, transfers of technology, exchanges of currency, and targeted restrictions such as those on bank accounts and international travel by specified leaders.

Many argue that sanctions failed in Iraq and Libya, but the accuracy of that assessment depends upon when the judgment was made. It looked as if sanctions fell short in Iraq because Saddam wouldn't allow UN inspectors full scope. But it was plain after the war that the sanctions had deeply wounded Iraq's economy and probably contributed to Saddam's secret decision to destroy his weapons of mass destruction (WMDs). The Libyan leader Qaddafi also appeared to have avoided the punishing effects of the sanctions, in good measure because he continued to sell oil. But even this oil-rich adversary was trailing economically, experiencing political jitters, and beginning to worry about his future. Perhaps the sanctions worked better than their critics have believed.

Economic sanctions have had very little effect on major powers such as the Soviet Union and communist China. Nor did these states, by the way, make major foreign policy concessions in return when the United States ultimately lifted the sanctions and permitted them to enter the global economy. Economic sanctions have no history of bringing down major powers or compelling them to change the course of their policies. In fact, even a number of small countries with efficient dictators such as Cuba and Burma have successfully resisted such economic pressure.

But American political leaders from presidents to candidates for national office uniformly have shown a great fondness for economic sanctions. They represent a way of appearing tough without going to war. For this reason, sanctions have become a favorite weapon in the American arsenal of power. It is this particular legislative pen they all grab when they can't figure out what else to do, but know they must do something.

For sanctions to have a good chance of success, four conditions appear to be necessary, if not always sufficient. First, the targeted country must be economically dependent on the sanctioning countries. Second, the sanctions must be multilateral so that alternative suppliers will be difficult to find. Third, the sanctioning countries must have the will, skill, and patience to keep the sanctions in effect over a long period of time. Sanctions are difficult to sustain in the face of inevitable cries that they hurt "the people" without harming the dictators, and charges that other countries are taking competitive advantage. Fourth, those imposing sanctions cannot make demands that they know the targeted leaders could never accept. Thus, if they seek to persuade a dictator like Castro or Saddam to give up power, sanctions will fail. If the sanctions ask leaders to give up something they might be able to afford, prospects are better. Thus South African whites decided they could better afford black political enfranchisement than they could endure global economic isolation.

Some think that being a major source of a scarce resource such as oil and gas is the surest route to economic power. In fact, this case is not conclusive. Oil-rich countries, such as Saudi Arabia and Venezuela, can amass vast fortunes from the sale of their valuable oil, but it is far from clear that they have been able to gain serious economic or political concessions in return. The only oil-rich nation that seems able to exercise economic power in the political sphere is Russia, which possesses military and other sources of power in addition to oil.

The greatest economic power lies in trade and investment, where the profits are widest and biggest and the promises of gain are most attractive. From 1950 to 2005, global trade leaped from $400 billion to roughly $20 trillion in 2008. Put another way, trade expanded 50-fold since the founding of the General Agreement on Tariffs and Trade (GATT).

Policy literature abounds with advice on how to succeed in trade negotiations. But far less is written about how to use trade for strategic, political, or security goals. Indeed, trade officials don't like foreign and national security specialists meddling in their business. They believe that the health of the economy is vital in and of itself, and that maintenance of a liberal and open world economic system is so important in its own right that no other considerations should interfere. They argue that rather than tying trade up in strategic and political knots, the government should concentrate on fighting the real threats of protectionism and rise of exclusive regional trading blocs: that is, any measures that limit the global market forces. From a single and open world economy, they believe, most other good things will flow.

The United States trade power rests on the size and strength of the American market and its strong position within the World Trade Organization (WTO), the governing body that oversees all global trade. This size and strength are a blessing and a curse. They are a blessing because historically the United States has been big and strong enough to make concessions and thrive—nonetheless Washington has been in the driver's seat in terms of bargaining in trade matters with other countries. But they are also a curse because other countries tend to wait for Washington to offer generous access to the U.S. market in order to reach an agreement. To say "Washington decides" to grant such access is somewhat misleading. Rather, there is a complex domestic bargaining process involving the president, various congressional offices, and business associations— which often conflict with one another. Still, little can happen in any of the global trading rounds, such as the Doha round, unless and until Washington puts more than its share of chips on the table.

In addition, the United States has the de facto power to block entry into the WTO, the body that sets and enforces global trading rules. Since it's impossible for a nation to reap the full benefits of trade without being a member, the American power to block admission to WTO is formidable. Unlike trade talks, which tend to be a world unto themselves, bargaining over entry into WTO often spills into unrelated strategic or political questions. For example, the Clinton administration extracted vague "understandings" about human rights from China in return for acceding to Beijing's membership. Bush tried to extract promises from Russia—not to use oil to blackmail neighbors, not to backpedal on democracy, and to help the United States block Iran's nuclear efforts—in exchange for WTO membership. Moscow did not show much interest in the deal. In such circumstances, most presidents opt to gather whatever political concessions they can without pressing too hard, and move on.

When Washington tries to extract specific political or security concessions in return for trade concessions, it is usually unsuccessful. For example, while China successfully rebuffed U.S. attempts to link preferred trading status with the United States, known as most-favored-nation (MFN) status, to human rights, securing MFN without such concessions, it is now delivering on economic reforms promised as part of the deal to allow its admission to the WTO. Trade power seems to work best when used in trade negotiations, and when it is just allowed to sit there for others to bump into and think about as they contend with noneconomic issues.

Americans are the biggest investors in other countries, but the great bulk of this investment is in private rather than governmental funds. Most other countries, like Saudi Arabia and Russia, control key sectors of their economies and can employ investment power through their central banks or through new sovereign wealth funds—huge sums from export profits controlled wholly by governments. Many states now have gone into the investment business directly with so-called sovereign wealth funds. This phenomenon is new and its effects are still unclear, but a great deal of money is involved.

Washington's power is limited in the investment realm. It can attempt to persuade private firms such as banks to make or to avoid investments in particular countries, but the ultimate power of decision in most of these cases rests with private companies. Interestingly, a large number of American and foreign banks acceded to the Bush administration's pressure to cut off or reduce banking transactions with Iran. The United States can also make it easier or more difficult for private firms to invest abroad. It does this through its various export-support entities such as the Export-Import Bank and Overseas Private Investment Corporation (OPIC), which are empowered to grant or withhold low-interest loans and provide insurance.

The United States' foreign investment totals are substantial. Official development assistance by the U.S. government has averaged $23.7 billion annually for the last five years. By contrast, private direct investment abroad approximated $3.7 trillion in 2008. Recipients prize these investments, including the technological benefits that often attend them, as much as they do trade. No other country approaches American totals in trade and investments, and this fact is never far from the minds of leaders who realize they live in a world where economic power now surpasses military power. And the United States has the most of both.

TALK OF ECONOMIC ISSUES such as these—investment, trade, economic sanctions, the value of the dollar, oil and gas prices—was rarely heard when I entered the foreign policy field decades ago. Military and economic aid were about as far as my colleagues ventured into questions of money. The Cold War was clogging up the world economy, and security issues hogged most of the attention, as they had throughout most of history. Besides, Americans didn't have to worry a great deal about money. While the American economy had its ups and downs, it was mostly up. We could afford to defend the world, even to pay others to defend it, which we did. We could afford to leave economics to the economists, businesspeople, and financiers, and by and large, we did.

But two things occurred in the last twenty years or so that undid this separation and ended the complacency. One was the opening up of borders and the explosion in global economic transactions, principally as a result of the end of the Cold War. Money issues now mattered much more to every nation. The other was relative uncertainty and growing concern about the health and strength of the American economy, even more so after the 2008 crisis. This, in turn, reopened questions about the relationship of our economy to our national security and foreign policy.

But the renewed discussion has not gone nearly far enough. We still make decisions about war and peace with little or no reference to America's economic needs. It's as if we still haven't internalized the basic truism stressed repeatedly by one of America's founders, Alexander Hamilton.

In his 1791 treatise "Report on Manufactures," Hamilton argued that a strong economy was the sine qua non of a strong military, and that both were absolutely necessary to protect and advance the nation's interests. In careful oldspeak he wrote: "Not only the wealth, but the independence and security of a [c]ountry appear[ed] to be materially connected with the prosperity of manufactures."

American leaders are only beginning to think again about Hamilton's point, about the centrality of economics to national security—and have been even slower to act upon it. The decisions they eventually take will have more effect than anything else on our military capability and our foreign policy power.

Meantime, America goes on about its foreign policy business, and for all the question marks that remain about foreign economic power and how to use it, some dos and don'ts "seem clear," as those given to ambiguity would say.

RULES FOR ECONOMIC POWER

Rule 1: America's power in the world depends on one factor above all—the strength and vibrancy of the American economy. Keeping it that way should be your absolute number-one policy priority.

Our roughly $14 trillion economy sustains your extensive military, diplomatic, and economic superiority and also our well-being and stability at home. It has been our central strategic asset for almost a century, and it is now vulnerable. Your first move should be to fix it.

Real fixes always go to fundamentals: good and competitive public education, the reconstruction of first-class infrastructure, lessening dependence on critical imports such as energy, governmental regulation that restores public confidence without undermining business initiatives, and reducing the federal debt and trade deficit. This is a very tall, but accomplishable, set of goals.

Rule 2: Economic power functions best when you permit it to proceed slowly, allowing it to act like the tide. This means you should use it more for medium- and longer-term strategic ends than for short-term tactical pressures.

A strong economy speaks for itself; it doesn't have to be wielded as a sword. It stands as a fact that all countries must recognize and accommodate, if they choose to participate in the globalized future. Let the leaders of other nations dream and worry about being a part of this, ultimately allowing them to coerce themselves.

Rule 3: Your economic power will be most effective if you use it principally for economic ends.

Make trade and investment decisions mainly on their own merits. It's too hard to try to swap economic apples for political and strategic oranges or exchange tangible money for gossamer strategic and political gains.

Rule 4: If you feel you must deploy economic power to address strategic and political matters, make sure the latter are stated as specifically and concretely as possible. Otherwise you will be likely to give something for nothing.

And don't wield instruments such as economic sanctions and aid unless you have good reason to be confident that they will succeed. You'll undermine our economic power if you use them just to get domestic critics off your back.

Rule 5: If you really want your economic power to work, set up a system to combine the economic, defense, and foreign policy departments of the executive branch. There must be a White House coordinator, and you have to back up that person.

You might even do your successors and your country a real favor and fund university programs on political economy. During the Cold War, Congress passed the National Defense Education Act of 1958, under which thousands of American graduate students had close to full tuition support for the studies essential to our success in the Cold War. Today we need the same level of federal encouragement for the study of political economy. If you truly believe that economic power is now the principal coin of the international power realm, this program will equip your successors to do that critical job well.

CHAPTER 10

STAGE-SETTING POWER

Soft-power instruments like persuasion, good values, and leader-ship won't—by themselves—cause foreign leaders to do your bidding where their own interests dictate otherwise. But they can help make foreign leaders more receptive to your real power: the pressure of your carrots and sticks.

Some of the most unpleasant clashes I've had in recent years have been over the question of whether soft power is really power. To me, soft power is foreplay, not the real thing. It's very important, but it isn't power, and it doesn't get the job done by itself.

The arguments I've had over soft power are usually with liberals, but more recently with conservatives as well. They think I'm being picky. I think they've all been so frustrated with President Bush's gratuitously bad treatment of other states that they've come to confuse behaving better and more sensibly as an international citizen with power itself. They rightly see that blazing paths of righteous-ness and goodwill, demonstrating better understanding of others' cultures, and being truer to our own ideals can do good things. But they overlook the fact that these good acts alone will rarely cause leaders to alter their assessments of their own national interests or do what they don't want to do, which is what power is all about.

There's a price my colleagues pay for their confusion: They aren't able to fully exploit soft power in the manner in which it can be truly effective—as a stage setter or as a mood enhancer, as a means to increase receptivity before the application of pressure and coercion. In fact, they're so fixated on their self-defined soft-power tools that they scant the other key stage-setting measures—namely, how well America solves problems at home and how sensitively it frames and presents policies abroad.

To see what stage-setting or foundation-laying power can do, you first have to see what soft power can't do by itself, and then look at its beneficial effects when it is used correctly. George W. Bush's handling of the run-up to the second Iraq war illustrates the first phenomenon, and his management of war strategy in his last two years illuminates the second.

IMAGINE WHAT MIGHT HAVE happened if Bush had fulfilled his critics' dreams on the road to the invasion of Iraq in 2003: if he had shown respect and understanding of Muslims, given Muslim-Americans presidential medals, shepherded Palestinians and Israelis toward a common oasis, and maneuvered deftly and judiciously for approval by the United Nations to wage war against Saddam Hussein. Suppose further that Bush had proffered an olive branch to Baghdad, including a firm pledge not to oust Saddam from power. This bouquet of sensitivity, internationalist values, leadership, and persuasiveness would have posed the ultimate test of soft power, of what might have been had the wishes of soft-power advocates all come true.

To begin with, let's posit that Bush would have been a model of religious tolerance and understanding, a veritable Benjamin Franklin. Let's say he would have lauded the virtues of Islam and Arab culture. Saddam and his henchmen would have laughed it off. They were secularists and didn't much like Islamic influence on their terrain, and they courted Muslims only late in the game and as a propaganda move to counter Washington. Most Muslims worldwide might have cherished such praise, if they had found out about it and if they had believed in the speaker's sincerity. But what the soft-power people don't appear to understand is that Muslims simply don't like Americans on Muslim soil. They would have regarded the invasion as proof of bad American intentions—however sensitive American rhetoric had been leading up to the invasion.

Or let's ponder the benefits of conveying to Arabs and others America's love of freedom, equality, human rights, and democracy. Saddam's Arab neighbors would have either shuddered at such incendiary talk or simply ignored it. Obviously, many Arabs and Muslims share these values and would have been pleased to hear Americans reaffirm them. But there's no abundant evidence that majorities in Muslim and Arab countries long for anything resembling Western-style democracy, political equality for all citizens, and the bedrock values of religious tolerance and free speech, or that Islamic cultures are congenial to any of these ideas. As for the Arab and Muslim leaders whom Washington supports, they're not fans of these values. Few, if any, ever uttered a harsh word about Saddam for murdering and brutalizing his citizens; they didn't care. They railed against Saddam only when he tried to incite their citizens to rebel against them or when he invaded Kuwait. To many Muslim and Arab leaders, American values threaten their power. Only the Muslim Brotherhood, Hamas, and Hezbollah might like Washington to live up to its values, particularly its commitment to democracy, because they'd be the beneficiaries. To them, free elections mean their winning and taking power and ousting America's allies.

Soft-power advocates also pleaded with the Bush team to go the last mile for a UN endorsement of force and not begin the war until he got one. But Arab and Muslim leaders didn't want any UN resolution that favored a U.S. and international invasion of Iraq. If America could gain legitimacy from the UN to invade one sovereign Arab state, it could do so for others. China, Russia, and many African states harbored the same distaste for such a UN resolution, one that could return to haunt them.

Nor would Bush have carried the day at the UN or among most

of its constituent countries by emphasizing any more strongly than he did the dangers of weapons of mass destruction, particularly nuclear weapons, in the hands of Saddam Hussein. To begin with, most of these states were already convinced that Saddam had WMDs, and they still didn't back the American invasion. Eventually, Bush rounded up a few dozen states that, with the exception of the United Kingdom, sent mostly token numbers of troops to Iraq. Most states didn't even want to approve the maximum economic sanctions against Saddam, let alone military force, for fear of humanitarian and other repercussions.

All this, in turn, suggests that Bush could not have obtained the approval of the UN Security Council for the war if he had stood on his head. Most states, including many that sent troops out of loyalty to the United States, didn't regard Saddam as a major threat. Besides, they didn't want to do anything that might stir up their own Muslim populations. Nothing short of Saddam's invading Kuwait for a second time would have roused the Security Council into permitting a second war with the butcher of Baghdad.

By the way, the Security Council did approve seventeen anti-Iraq resolutions—repeat, seventeen. Most demanded that Saddam comply with inspections to prove he had abandoned WMDs, but not one contained a clear "or else" clause threatening military action. Only a cockeyed optimist could contend that had Bush waited longer, an eighteenth resolution would have generated an international blessing for war.

The much tougher question is what might have occurred if Bush had taken the ultimate soft-power step—publicly promising Saddam that if he complied with the seventeenth resolution, abandoned WMDs, and allowed comprehensive inspections, he could remain in power. Indeed Bush—or at least his secretary of state Colin Powell—made just such a suggestion as the day of the attack approached. But there's no evidence that Bush really would have gone through with this or—more important—that Saddam would have trusted Bush to make good on such a promise. And even if Bush had held back, Saddam certainly would have calculated that he couldn't hold on to his power under those circumstances. Remember that at this point, he was so stressed out by his vulnerability that he had already eliminated his WMDs, but was afraid to tell anyone.

I think it's fair to conclude that the vaunted soft-power moves would not have manufactured much additional support for Bush's Iraq policy. The issue of whether to take up arms against a sovereign state that had not attacked another was far too important to be resolved by being more persuasive and reasonable and living up to our values. The only states that made noticeable contributions to the American war effort were those few that shared American concern about Saddam's WMDs, such as the United Kingdom and Spain (for a while) and those that, like Poland, needed U.S. support in their own neighborhoods. None of this is to say that Bush shouldn't have availed himself of most items from the soft-power menu. At a minimum, it would have softened the subsequent anti-Americanism, which damaged U.S. policy worldwide.

If we jump ahead now six years to the end of the Bush administration and observe what Bush and the Iraqis accomplished in Iraq, even without soft-power inputs, we see that for the first time, after so many policy mistakes, and after the squandering of so many American and Iraqi lives and so much American treasure, the situation in Iraq is promising. It doesn't portend a stable democracy, but the situation is relatively quiet, American casualties are significantly down, and Iraqis are back on the streets in most provinces. This has provided the allure if not the substance of stability. All this has happened in the face of Abu Ghraib and other horror stories about American conduct, including some of the worst tales of American torture, combined with an insensitivity to Iraqi history and culture that continues to boggle my mind.

Events in Iraq took a turn for the better because the U.S. military changed its strategy from one of staying in its camps to one of putting boots on the streets, to a strategy of winning, holding, and building Iraqi society. It also helped somewhat that Bush sent more troops to Iraq. More important than these good fixes in military strategy were the changes that Washington finally made in its political strategy. Principally, the administration opened itself to common cause with former Sunni Arab enemies against their common enemy—al-Qaeda. Local Sunnis knew far better than we who the terrorists were. The result was that we jointly delivered the terrorists a heavy blow. At the same time, Bush finally agreed to a de facto cease-fire with the Shiite militias. Also, to everyone's astonishment, the hapless Prime Minister Nuri al-Maliki turned into Iraq's version of The Hulk and began to act decisively against disruptive Shiite militias, his coreligionists.

If anything turned around the war in Iraq, it was these changes in strategy and policy. They were so sound, so in tune with emerging Iraqi realities, that they helped establish America's new security strategy—in the face of the most unattractive American behavior earlier in the war.

SOFT POWER BY ITSELF could not have delivered more allies or public support before the 2003 Iraq War, nor were classic soft-power instruments essential for later improving the situation in Iraq. Soft power simply didn't work or matter on these two critical fronts. But it could have facilitated our exercise of power and enhanced general receptivity to our power. It could have prepared the ground for real power—pressure and coercion—by demonstrating that Americans took others seriously and were willing to hear their views and take their political problems into account. Power entails pushing people around, and the ones being pushed never like it. Applying the full spectrum of stage-setting measures permits the pushees to feel that they're being treated with respect, and that does help.

If the soft-power package is delivered as a prelude to the later application of power, it cloaks and cushions the ultimate blows of power. The domestic audience in the target countries and one's own partners don't notice or feel as much that they're being manhandled. Sensitive public diplomacy and shrewdly conceived policy help make being pressured more palatable both politically and psychologically.

. . . . .

THIS COLD AND LOGICAL presentation of the benefits of stage setting makes it sound easier than it is. In fact, it's difficult—an art that we can better understand if we return to the beginnings of soft power, which were not in the twenty-first century but rather in the awe-inspiring displays mounted by emperors' courts in premodern China, and in the propaganda machines of Hitler's Germany and Stalin's Soviet Union. In the United States, Americans charmingly referred to enterprises with similar purposes as education and information programs.

The emperors of China saw the connection between a state's reputation abroad and its prowess at home, but believed that prowess didn't entirely speak for itself. It had to be enhanced, artfully embroidered. They were masters of dressing themselves up for visiting foreigners and making them feel small and powerless in the presence of the emperor and his court. They were in the awe-producing business and understood that it was cheaper and probably more effective than war. They grasped a thing or two about how to arouse fear and respect with a show of their riches, with full displays of pomp and circumstance, topped off with the exotic custom of kowtowing. They saw the magical properties of reputation and propaganda.

Few will be surprised to learn, however, that the word and practice of propaganda originated in the West with the Catholic church. In 1622, after a particularly nasty period between Catholics and Protestants during the Thirty Years' War, Pope Gregory XV founded the Congregatio de Propaganda Fide, or Congregation for the Propagation of the Faith. This committee of cardinals oversaw the propagation of Roman Catholicism by missionaries to non-Catholic countries. "Propaganda" sprang from the Latin *propagare*, "to propagate, extend or spread." Nothing in its original meaning even hinted at the idea of selling a story, let alone purveying misleading information. But instinctively, the church understood what it needed to do to spread its version of the faith.

Centuries later, when the public relations industry was born early in the twentieth century, its practitioners were not embarrassed to refer to their activities as propaganda. After all, there was nothing shameful about being paid to disseminate truthful information. The Nazis and the Soviet communists appropriated the word without shame, as in their ministries of propaganda, but there the term eventually lost its neutrality and finally took on a clearly pejorative cast. The propaganda artists sold Nazi Germany as an unstoppable master nation led by a master race. Germany's neighbors recoiled at being labeled mongrel peoples, although they were impressed by Nazi uniforms, goose-stepping troops, and public extravaganzas. The Soviets offered themselves as revolutionaries to an international public yearning for saviors of the downtrodden. They would fight for the poor against the exploitive capitalists and capitalist nations. In their story line, the Communist Party would overthrow the exploiters. From the Nazis and the Soviet communists onward, propaganda had a bad name. It became synonymous with disinformation and lies.

Americans have been aware of the idea of propaganda since revolutionary days. But they weren't consciously cynical about using it; they just acted cynically when necessary. The American style was to tell the truth and feel great pride in doing so, and enlarge on it only when necessary to vanquish adversaries. The leaders of the American Revolution, of course, truly believed in freedom, equality, and independence, but also recognized the need to sell their revolution with a very moderate, unthreatening face to Europeans whose support was essential to America's independence and national survival.

A century and a half later, in the wake of World War II, a new generation grappled with the issue in Congress, and came up with a clever new way to conduct propaganda campaigns on behalf of a democracy. The Smith-Mundt Act of 1948, also known as the U.S. Informational and Educational Exchange Act, eschewed the term "propaganda" to describe the information it authorized for foreign audiences but prohibited from being disseminated within the United States. In other words, the act authorized the U.S. government to say what it had to say abroad to defeat the Soviets, but shielded American citizens from this rhetoric.

These educational and informational programs evolved into the U.S. Information Agency (USIA), which sponsored overseas radio networks; publications; exchanges of students, professors, and others; and cultural programs across the globe. Many of these programs served the country well during the Cold War, but the whole effort lost political support and funding after the demise of the Soviet Union.

What did survive was eventually subsumed under the label "public diplomacy," a phrase attributed earlier in 1965 to Dean Edmund Gullion of the Fletcher School of Law and Diplomacy at Tufts University. As explained by Gullion, this kind of diplomacy "deals with the influence of public attitudes on the formation and execution of foreign policies. It encompasses dimensions of international relations beyond traditional diplomacy; the cultivation by governments of public opinion in other countries; the interaction of private groups and interests in one country with those of another; the reporting of foreign affairs and its impact on policy; communication between those whose job is communication, as between diplomats and foreign correspondents; and the processes of intercultural communications."

Gullion's definition was silent on the issue of whether public diplomacy needed to be truthful, a question that Edward R. Murrow, who was then the director of the USIA, addressed head-on in his congressional testimony in May 1963, when he said, "American traditions and the American ethic require us to be truthful, but the most important reason is that truth is the best propaganda and lies are the worst. To be persuasive we must be believable; to be believable we must be credible; to be credible we must be truthful. It is as simple as that." Would that it were.

I THINK IT'S FAIR to say that the U.S. government did a much better job of selling America when it didn't have to sell America.

That was the America of the first two centuries of its existence, up to the Vietnam War. For those centuries, America was the "new world" where everyone could start equal, "the city upon a hill" of religious tolerance in a world of intolerance, the land of economic opportunity, and later the savior of freedom from tyranny in the first and second world wars. Foreigners flocked to its shores as immigrants and foreign students came to its great universities. Its culture was irresistible, and still is, though many now condemn it as well for being materialistic and spiritually empty. This America did not have to be sold. This America was admired and powerful, an emotional magnet.

It used to be virtually axiomatic that Americans were the good guys in the world. Then came the Vietnam War, along with condemnations of the ugly—that is, the ignorant and arrogant—American. From then on, the regard for America was mixed.

Most began to hold the United States to higher standards than other states, mainly because Americans laid claim to those higher standards. By contrast, the Chinese, who made no assertions of moral superiority, received few charges of hypocrisy. Worse, Washington started to lose ground in world opinion to terrorists and ideological extremists. Opinion polls in countries like Pakistan—where, admittedly, citizens are even less likely than Americans to share their true beliefs with a total stranger—found that Osama bin Laden had more admirers than George W. Bush. Partly, this situation sprang from widespread knowledge of and disgust over Abu Ghraib, the Iraq War, and U.S. support for Israel. And because the United States was seen as the globalizer in chief, it was also blamed for economic inequalities around the world.

Much of this criticism was unfair or grossly unbalanced. Underlying it all, albeit to a lesser degree, America is still admired in many ways around the globe for all the things it used to stand for—and still does. But anti-Americanism is now woven into the texture of international affairs.

Some believe this anti-Americanism is only skin-deep or related specifically to George W. Bush and will largely evaporate now that he has departed from the Oval Office. They point to the extremely high anti-Americanism of the Vietnam War days, which mostly dissipated in a few years. Obama moved quickly to alter America's reputation among elites and general populations worldwide. In speech after speech, he showed them he wasn't stuck in a narrow American understanding of their problems but actually saw these problems from the local point of view. And he did this without making negative concessions.

IT IS TYPICALLY AND charmingly American to reduce the sensitive process of combating these problems to salesmanship, or as it's now dubbed, public diplomacy. Not surprisingly, Bush first appointed a Madison Avenue specialist to the job, and then a friend and political adviser whose main expertise was in Texas politics.

Certainly there's salesmanship in public diplomacy, but if such diplomacy is to be rightly practiced, it has to be conceived of as much broader and more complex than merely marketing. It has to be thought of as stage-setting power and understood as having three main components: (1) The soft-power package of values—leadership and the power of example to attract support. (2) A reputation as a nation able to solve its own domestic problems and thus worthy of inspiring both awe and confidence abroad. (3) Foreign policies and diplomacy to demonstrate sensitivity to and appreciation of others.

All these efforts, by the way, must be country- and region-specific and sensitive. When in Moscow, treat the new Russian czars like powerful partners and jointly strategize with them. Show respect for the Chinese because of both their rich history and their legitimate claims to esteem based on their present-day economy. When in Iraq, always let leaders there see that you understand it's their country. For Europeans, evoke international law and the UN, and then get down to business. Like us, foreigners have their cultural tics, but their opinion leaders are sophisticated and knowledgeable. They judge Washington not on its salesmanship, but on what it is actually doing to and for them, and how.

More important than soft power in persuading others to be receptive to American power is America's reputation for wealth, dynamism, effectiveness, and problem solving. Others take our measure all the time. Is America still the land of greatest opportunity, and does its economy remain innovative? Is it a society where different kinds of people of different religions can find tolerance? Is it politically fragmented and stalemated, or can its government still effectively solve major problems such as the costs and quality of medical care and public education?

The ancient Chinese believed that in exercising power it was critical to sell your domestic prowess. They were on to something essential. The successes of a society, an economy, and a government carry like sound waves and have the additional virtue of more or less speaking for themselves. Most around the world can clearly hear whether a nation is performing well and on all cylinders.

The reputation of the United States is in flux. It still benefits from being the victor of the Cold War, but other factors like the Iraq War have sullied it. The U.S. economy remains number one, but it has experienced rocky times, especially with the 2008 financial crisis, and is less competitive now than in the recent past. Equally clearly, America's reputation as the country that gets things done is waning. And with this decline, Washington's words lose weight abroad.

There's a story, perhaps apocryphal, about Napoleon, one of the ultimate men of power, and his storied minister Charles-Maurice de Talleyrand-Périgord, one of history's premier judges of power. Napoleon was given to dropping his hat here and there for those nearby to pick up. Talleyrand was a particularly obliging retriever of the hat—until after Waterloo, when Napoleon dropped it and Talleyrand pretended not to notice.

Nations around the world are not as attentive to our every move as Talleyrand was to Napoleon's, or as they were in the past. When we were comparatively wealthier, and when the dollar was the undoubted and stable reserve currency, they thought we really knew everything. Our power and standing gave us the high ground in disputes without our having to argue for it.

No selling skills can replace the slippage in America's reputation. We can hope that American leaders will make hard and smart decisions to restore an America that is both good and strong. For all the anti-Americanism, you can feel that most people in most countries would like America to succeed as a multicultural land of opportunity and a global guardian of human rights—whatever they might say.

Meanwhile, there's only one way to make up some of this lost ground—through the third ingredient of stage-setting power, namely, better policies at home and abroad. The aim of policy is to explain how you propose to solve problems. Done well, policy permits you to build support or blunt opposition. It supplies the explanation of why you chose your course and not the alternatives. Policy is where you draw the picture of what's in the solution for others and what they're in for if they oppose you. Policy must be firm in its judgments, yet sensitive to the culture and interests of the intended audience.

None of this is to say that Washington should bow to foreign demands or pander to hostile foreign opinion surveys. American interests have to come before popularity in the polls. Policy that is sensitive and skillful, however, can boost both. Polls do tell which way the wind is blowing, and what foreign leaders will have to overcome to comply with our wishes. So they shouldn't be dismissed. Not tending to these voices and the politics they engender guarantees just one outcome—stronger resistance to the application of U.S. power, and the need for more power down the line to surmount the increased resistance.

Why do these obvious points even have to be stated? Because only in America would there be strong doubts and suspicions about the commonsense practice of courting foreign peoples and groups to make them more cooperative or less resistant to American power. Because only in America do leaders pay so little heed to the links between stage setting, public diplomacy, and policy.

Policy is the principal means of providing political cover to foreign leaders who don't want to be seen as knuckling under to raw American power. Applied intelligently, it helps foreign leaders explain to their constituencies why they're submitting to American demands and what they're receiving in return. Sensitized policy allows them to say they are doing this in return for that, or that their country's point of view is being respected, and that the United States is seeking a fair resolution.

The costs of gratuitously and carelessly ignoring these sensibilities have been high. In March 2003, to cite but one example, the Turkish parliament refused to approve the transit of U.S. troops through their country into Iraq. Washington's offers of billions of dollars in aid weren't sufficient to appease the opposition. The Bush team had neglected to assuage Turkish fears that a war in Iraq would strengthen the base of rebel Kurds in Iraqi Kurdistan. The Bush team also made an insufficient effort to counter charges of American prejudice against Muslims, or to patiently remind Turkey of Washington's help to Muslims in Bosnia and Saddam's punishment of them in Iraq. If these points had been raised, the Turkish parliament might well have been of greater help to American forces.

George H. W. Bush and Bill Clinton both took pains to express to leaders of the newly independent former Soviet republics that American power stood by them. These messages were essential to persuading them to return the nuclear weapons on their soil to Russia. George W. Bush was the first U.S. president to state publicly and explicitly that Washington would support an independent Palestinian state. Palestinian leaders needed that assurance to be in a position to make future compromises. Unfortunately, Bush didn't have much of a follow-up policy. Obama's message to the world when he took office was "I'm ready to hear you."

Finally, stage-setting power permits the president to exploit latent pro-American feelings in most parts of the world. For all of America's new woes, most foreign opinion leaders still realize that the United States is the country with the most power and ability to help them solve problems. Most also see—for all their surface anti-Americanism—that the United States is probably still the most well-disposed of all the major powers to be of help to them. Sometimes, they worry that the American cure might be worse than the disease, but they usually seek Washington's help anyway. Most recognize, as well, that for all Washington's clumsiness, it is more benign in its intentions than the others. Those in need rarely run to Moscow or Beijing, whose cynicism exceeds our clumsiness. America's advantages are dwindling but remain real and keep doors open to American involvement and power.

Policy, the key ingredient of stage-setting power, works best when complemented by a strong reputation for national prowess plus soft power. Presidents need all three elements in play along with real power to give themselves a chance at advancing negotiations and dampening conflicts. All have to be meshed, and there are a few straightforward rules for orchestrating this process.

RULES FOR STAGE-SETTING POWER

Rule 1: Don't think of stage-setting power as just soft power or marketing dressed up as public diplomacy. It is a three-part package of soft power, reputation as a successful nation, and good policy.

The soft-power tools do help foster receptivity and overcome opposition. Reputation for national prowess in solving internal problems establishes credibility; it shows that you know how to get things done at home. Policy explains your interests and theirs in making a settlement.

Rule 2: Think of stage setters as facilitators of your power, but not as power itself. These tools provide, not actual incentives to compromise, but rather the ammunition for foreign leaders to justify their compromises to their home audiences.

Power facilitators can make foreign leaders more receptive or less opposed to your power. Applied skillfully, they can show foreign leaders how to meet your wishes and survive politically in their own countries.

Rule 3: Don't try to substitute stage setters for power itself, because they are almost never sufficient to induce foreign leaders

to risk significant compromises. Stage-setting power can bring
leaders to water, but it can't make them drink.

Think about it this way: Would you make important concessions to the United States if you were running Iraq and your
only public rationales were "We share American values," "The
American position is really good for our long-term interests," and
"America's a great country." In the end, only strong carrots and
sticks would move you. And frustratingly, even doing it all right
doesn't usually work initially, and you have to start the process
again and again.

Rule 4: If you really want to be serious about stage-setting
power, you must fully integrate it with the policy formulation
process. If the initial policy lacks sensitivity and doesn't help the
other guy do what you want, it won't work and will make it harder
to recover for the next round.

To decide on a policy and only afterward figure out the public
diplomacy is nonsensical. By then it's too late. There's no sense
in bugling for the public diplomacy cavalry after the settlers have
already been wiped out. If the legions of recent devotees to public
diplomacy are interested in more than posturing, their first fight
must be to ensure that experts sit with policymakers from the
outset. Everyone has jumped onto the public diplomacy band-
wagon, but few senior officials in government have been willing to
do what's necessary to practice the art seriously.

Part III

Policy and Power

CHAPTER 11

FOREIGN POLICY POWER

Policy is where you bring American power to life by words and deeds.
Policy is where you stitch everything together—the intelligence, the
politics, the strategy, the messy inconsistencies—define the problem,
and show how you will solve it.

Good policy is often the thin blue line between the possibility of success and the certainty of tragedy. Take those few weeks of combat and the early months of the U.S. occupation in Iraq in 2003. The Bush administration felt either that it didn't need a policy for what Pentagon bosses judged would be a short stay, or that we would simply transform Iraq as General MacArthur democratized postwar Japan. I played a brief and modest role with key figures in the White House in a failed effort to put a realistic policy in place before it was too late.

It didn't take long into the war for Middle East experts to see that American victory on the battlefield would be the easiest part of our efforts in Iraq. With Saddam out of the picture, the ages-old centrifugal forces in the country would probably implode. I had been listening to these mounting lamentations in meetings at the Council on Foreign Relations, of which I was president. So, I phoned Condoleezza Rice, Bush's national security adviser, and her deputy Stephen Hadley. I had known both for decades in the Soviet and arms control business, both were members of the Council, and I respected them.

I made the case about the problems, and neither balked. I said perhaps the Council could partner with a couple of other think tanks, collect a wide range of experts, lay out the issues and the facts concerning these issues, and simply list policy options for each issue plus options for an overall policy. I said I understood the White House would not want a finished product, which might cause wrangling and unfortunate news stories about policy differences, and stressed that we'd therefore also forgo recommendations. They accepted this approach, and we agreed that the Council would partner with the conservative American Enterprise Institute headed by Christopher DeMuth and the moderate Center for Strategic and International Studies run by John Hamre.

Rice convened us all in her corner office at the White House a week later and asked me to make my pitch for this study, and when I finished, she supported the plan. Then, DeMuth asked, "Does Karl Rove (the senior political adviser) know about this? Does the president? Because I don't think they would approve. It sounds like a nation-building exercise, and they—and you yourself, Condi—have opposed such foolish Clintonesque policies time and again."

Rice responded that the study would not make recommendations on policy, but would lay out the facts and policy choices. She startled me with her honesty by adding that she and Steve just didn't have the time, with all the daily decisions that had to be made, to oversee a similar effort inside the administration and that, in any event, we could gather together people with unmatchable expertise. DeMuth stuck to his guns. A few days later, Hadley called and without explanation said, "I don't think we're going to be able to do this. Sorry."

I established a Council group to plow ahead on Iraq policy on its own under the leadership of two eminent former diplomats, Frank G. Wisner and Edward Djerejian. Their report anticipated virtually all the problems the United States was about to confront and offered sensible policies for dealing with them. It received few notices. It wasn't until almost five years later that the Bush administration, facing the prospect of defeat, finally grasped Iraqi realities and fashioned approaches that turned events in a promising direction.

You may well wonder how this could happen and whether it was an aberration. Alas for our country, it is the rule, not the exception.

The fault lies not in the esoteric nature of policymaking itself. Rather, presidents and their senior advisers are not used to doing it and can't wedge themselves out of the daily grind of press and congressional demands. And they don't make the time. Policymaking takes time, and it really requires the kind of gut-wrenching discussions about fundamentals that most people just don't find comfortable. By saying what you believe to be important, you're also revealing what you think is sacrificial, what's worth fighting for and what's not. You will be hounded one way or the other. The process calls on you to decide whether you want to double down in a tough situation and stay the course or fold and seek another path. And everyone will be told about your position. There's no place to hide politically.

It's much easier to make daily decisions on how to answer press queries or whether to ask Congress for more money than to face and think through the basic policy questions. But unless you're willing to drown in your own pool—Iraq or Vietnam—it is essential to grapple with both overall and particular policy issues before the inevitable troubles erupt. This process is the only one that will compel you to evaluate whether your ends are wise and achievable, and your means sufficient unto those ends. Very simple, yes, but again, we very rarely do it.

THE STARTING POINT FOR a new U.S. foreign policy is to fix as precisely as possible on overall goals. America can't escape its global interests and involvements. America's interests and goals must thus be broad yet not bloated—always attentive to distinctions between the essential and the merely desirable.

Today, the United States has seven inescapable goals.

First, contain and then diminish the threat to America and its allies from international terrorists. We have to understand this as a long-term goal, mainly revolving around constructing intelligence, police, and judicial networks worldwide. But this goal will also require tough actions, including invading a country that harbors or sponsors terrorists.

Second, prevent the spread of weapons of mass destruction to states and especially to terrorists. If states nevertheless acquire WMDs, particularly nuclear weapons, deter them with the threat of decisive retaliation if their weapons should threaten or strike American or friendly territory. Leave no doubt that if a state provides WMDs to terrorists, the United States would consider this an act of war.

Third, prevent any state or group of states from organizing the power and resources of Europe and Asia against the United States. This objective is absolutely not designed to rationalize a new Cold War with Russia and China, as some already advocate. Rather, it is a reminder to American leaders that they must play the international power game on these critical continents to guard against strategically unfavorable trends and bargaining situations.

Fourth, promote freer international economic transactions, especially trade and investments. Unfortunately, Americans are once again challenging this goal, and its immediate future is in question. Many other states are now more competitive and pose threats to American jobs. For most of American history, a solid majority has had the capability and confidence to compete and prevail. A freer world economy also necessitates a dominant American navy to maintain freedom of the seas.

Fifth, significantly reduce dependence on foreign oil and gas within a decade, and in particular, reduce to as low as possible dependence on Middle East oil. The United States cannot afford the cash outflow, or the demands to protect questionable allies in the region. Cutting dependency on hydrocarbons also helps the environment.

Sixth, protect against global environment and health threats consistent with the principles of mutual state responsibility and economic feasibility. But recognize that we've reached the point where economic interests must bow to some degree to environmental and health imperatives, when a good case is made.

Seventh, remain faithful to and promote long-term efforts to develop the rule of law, democratic institutions, democracy, and respect for human rights, but be ever mindful of the substantial obstacles to establishing genuine democracies and to the costs of rushing this complicated evolutionary process. Focus principally on laying the foundations for democracy in civil society and human rights.

Some prefer to state these inescapable goals in more grandiose terms, whether it be a commitment to take certain military actions or to fight global warming whatever the costs. Still others insist on ranking objectives so that military ones are on top of environmental ones. All of these objectives, however, are of first-order importance. The best approach is probably just to hang them all on the wall, address them all, and deal most with what you must. One point to register is the connection between almost all these goals and the health of the American government and economy. Continuing to meet any of these goals, let alone the whole roster, is premised on an America that remains not only strong but also almost as strong relative to others as it is today.

The next step is to bring the inescapable goals to earth by tying them to relevant plans and means. Exactly what power can the United States draw upon to advance these goals? Every country's power is rooted in its resources and position in international affairs. In a mechanical sense, U.S. power comes from the carrots and sticks it can deploy relative to others in specific cases, plus relevant judgments on exactly what each party is likely to do the other. The United States has the most disposable resources of any nation and is better positioned than anyone to inflict pleasure or pain. But none of this bestows on the United States the power to solve or perhaps even successfully manage a major international problem on its own. Unilateral power and the power to dictate and dominate haven't been around for many decades and are gone with the twenty-first-century winds.

What superiority in resources and international position give to the United States is the unique power to lead. Only Washington can fashion coalitions both to stop aggressors and to solve major problems. No other state or group of states can assume this role, and the other states know this well.

But for Washington to grasp the potential in this unique role, it has to think of leadership power not as ordering power, but as galvanizing power. Galvanizing power is the power to trigger and shape coalitions, and set their directions. Without these coalitions, neither the United States nor its coalition partners could solve problems—and our partners are very aware of this. We are the indispensable leader and they are the indispensable partners. Even the Chinese are happy to see us lead—at least for the next ten years or longer before they near economic parity with us.

The condition of mutual indispensability is not an end in itself, but the basis on which to build the power coalitions necessary to solve those problems. To create those coalitions and then to wheel them into action must be the center of U.S. foreign policy. America's attendant diplomatic power must be geared to this process as well.

Here, too, the United States has to rethink and redefine diplomatic power. Instead of the familiar threatening power, diplomatic power now has to be "leaning" power. This power will convey the weight of U.S. resources without the side effects that normally generate opposition to meeting American demands. Public policy statements should be firm expressions of American interests and should mainly brandish carrots. Sticks are better bared in private. If the carrots and sticks truly matter to the target state, its leaders will feel the weight of American diplomacy.

It takes time, especially today when all leaders have their domestic political traps and snares. Presidents (and cable TV) seem to abhor patience. But the passage of time allows foreign leaders to convince themselves and their constituents that their prospective concessions are mutual compromises and that their compromises are triumphs. Leave it to them to work out. The process will always be plagued by the tug between political pressures for answers and results and the inevitably slow pace of diplomacy.

As DISCUSSED TO THIS point, U.S. foreign policy encompasses America's overall policy, goals, and means to meet its global concerns and opportunities. It's important to start at that grand level, and to deal with as many issues as possible at that level. For example, we could aim to have just a single universal policy to combat nuclear proliferation or advance trade. Such a policy would make it simpler to hold all countries to the same standards. But international life doesn't operate that way.

Most foreign policy is made toward one country or region, and at that point, the choices become stickier. In the land of the particular dwell trade-offs and dilemmas that are usually the hardest dimension of the policy process. These dilemmas mark every tough problem, and leaders must resolve them up front or have their power work at cross-purposes and cancel itself out. Policy whispers in their ears, "Hang on to all your options and goals." Power whispers in their ears, "Choose."

Once leaders jump or are tossed into the frying pan of a country or region, be it Vietnam or Iraq or Pakistan, they invariably confront horrible choices. Each choice is a tiger, biding its time while leaders deliberate. Leaders must choose whether to be eaten by one tiger if they do or the other tiger if they don't. The common practice is to straddle both tigers at once to avoid the jaws of either. Leaders fear choosing and making a big mistake. They forget that only pundits and foreign policy experts can ride both tigers. Alas, it's not the pundits and experts who get eaten.

When faced with most dilemmas, presidents must either choose or lose. It's fine to try prioritizing the taming game, but pursuing both tigers, even in turn, is generally unattainable and drains the power to tame the more dangerous tiger. Sometimes, the choices are truly perplexing; at other times, they seem more perplexing than they really are. Presidents should examine them at the very outset of troubles in order to organize and prioritize their power, not squander it.

Typically, the first dilemma is to decide whether the matter is vital or just another among dozens of important issues. There is only one method to address this dilemma without incurring maximum political wrath, and that is, early in a presidency, to establish a short list of vital interests and goals and sell it politically. At the same time, presidents also have to provide assurances that they will tend to the second-order issues in some plausible manner. The Latin American constituency in the United States wants serious attention paid to neutralizing President Hugo Chávez of Venezuela and creating a new trade relationship with a growing economic power like Brazil. President Obama has to be prominently involved in the policy trade-offs here, but should then turn the mission over to the secretary of state or the secretary of the treasury and then an assistant secretary. On a compelling humanitarian issue like the ongoing killings in the Darfur region of Sudan, President Obama has to publicly set the course, maybe appoint a special envoy, and then wheel himself in only for critical phone calls to his counterparts or to press Congress to pass relevant legislation.

A second dilemma is choosing, not between the vital and the important, but between two virtually vital interests. Presidents always want both, but rarely can manage that.

In 2006, Bush tried both to wheel India into a tacit strategic pairing against China and to keep faith with his tough anti–nuclear proliferation policy. He would bring India to America's side by acceding to its demand not to make the nuclear facilities run by the Indian military subject to international inspection. The nonproliferation treaty requires inspection of all such facilities. The upshot was that the Indian government publicly sidestepped taking any anti-China stand, and the anti-proliferation regime was fundamentally weakened.

Regarding recent policy toward Russia, Bush had to choose between courting Moscow to gain its critical help against Iran's nuclear efforts, on the one hand, and pushing ahead with the deployment of

U.S. antimissile systems in Eastern Europe, on the other. He tried to do both, apparently realizing too late that Russia's help on Iran was far more important than the anti-missile deployment. By that time, the prospective anti-missile deployment had angered Moscow and blocked cooperation elsewhere. It would have been easy to defer the anti-missile decision, since our technology wasn't ready, and Iran, by U.S. intelligence estimates, wouldn't be either eager or able to fire off missiles in the European direction in any event. Obama seems to have accepted this judgment.

The ultimate example of strategic loggerheads, one that surfaces regularly, is the choice between reshaping the world and protecting particular American vital interests. Democracy's advocates always cry out that American security will forever hang on the whims of dictators and terrorists unless Washington democratizes or conquers the world. Bush topped all previous efforts in this direction—save Woodrow Wilson's—by indicating that America would seek to transform Iraq into a free-market, democratic paradise and press our other friends in the region to convert as well. He and Condi Rice triumphantly announced the transformational plan, and a year later, effectively and embarrassingly returned to business as usual. Bush and Rice gained nothing in the propaganda sphere and enormously annoyed America's allies, such as Egypt and Saudi Arabia.

A third dilemma is to choose between changing a regime and changing the regime's offending policies and behaviors. We must alter not only what they say but also what they do. Washington managed to cashier governments in relatively weak countries like Guatemala and Iran during the 1950s and 1960s. In the 1970s, and later in the Clinton and Bush administrations, there was a lot of loose and popular talk about regime change. Both presidents flaunted this rhetoric and initiated overt and covert programs to dump the dictators in Libya, Iran, Iraq, Afghanistan, and North Korea. Bush managed to topple the regimes in Iraq and Afghanistan—by invading them. Otherwise, it's fair to say that the idea of toppling regimes has accomplished only a temporary warm glow among hard-liners, and a uniting of the opposition abroad against the American bully.

The alternative—maneuvering for desired policy and behavioral alterations—is no easy matter, either. The trickiest judgments are whether the nasty regime will compromise, whether the compromises will be sufficient and real, and whether promises will be kept. Doubts on any of these counts invariably derailed even exploratory talks.

In a number of instances, such as with Libya and recently North Korea, presidents have taken gambles. Most have paid off, in whole or in part. Situations change; the untrustworthy can discover reason and flexibility when their internal power is challenged and their exhaustion exceeds their hatred. Letting adversaries whiff potential benefits can also alter their calculations. During and even after the process, the military option need not be shelved.

Another frequent dilemma arises when there seems to be little hope for policy change. Leaders then have to choose between taking even bolder and costlier actions to alter that policy and simply containing the problem. Specifically, say a troublesome state continues its nuclear programs or backing of terrorists. What then? This was precisely the dilemma that confronted Clinton and Bush regarding Iran, North Korea, and at one time Iraq as well. In Iraq, Bush rejected containment from day one. He and his advisers simply assumed that the waiting game of containment would only permit Saddam to develop weapons of mass destruction secretly and become a deadlier foe later on. Bush didn't seriously consider the United States' ability to deter Iraq from employing these weapons by threatening devastating retaliation. He rejected deterrence even though there was reason to believe Saddam might heed it. After all, Saddam had suffered defeat in the First Gulf War without firing off his chemical weapons.

As for the blossoming nuclear threats from Pyongyang and Tehran, Bush initially appeared to believe—without any evidence—that U.S. military threats would frighten these states into abandoning their nuclear programs. But after they repeatedly sped through his red lights, Bush increasingly leaned on joint diplomacy with Asians on North Korea and joint pressures with Europeans, Russia, and China on Iran. In both cases, he toned down threats. Meantime, he had badly wasted his credibility.

Another dilemma turns on whether or not to tie the solution of one policy problem to another unrelated or indirectly related policy problem. The theory behind linking is that this adds bargaining leverage to both issues. But the downside is that linkage can also complicate the resolution of both problems, and thus kill or delay sensible compromise.

Knitting issues together in this manner is an ancient and honorable practice, but Nixon and Kissinger coined the term "linkage." They employed it most famously in bargaining with the Soviet Union. Specifically, they told Moscow that reaching compromises on various arms control negotiations (which presumably Moscow needed more than the United States) would depend on Moscow's removing the threat to Berlin, easing tensions in the Middle East, helping end the war in Vietnam, and curtailing its support for wars of liberation in places such as Africa and Central America. Moscow refused this linkage and it didn't work.

Linkage is sometimes foisted on presidents when one of the issues concerns human rights. Proponents of linkage reckon that dictators aren't likely to bow down on human rights questions alone, so they look for another issue to provide added leverage. Typically, that lever is money. For example, Egypt commits rights abuses and doesn't want to fix them, but it continues to want U.S. economic aid. The linkers would make the aid dependent on rights reforms. But no president has been prepared to try this linkage; instead, presidents have reasoned that President Mubarak is too important an ally to risk undermining and his economic dependence is not as large as it used to be.

Iran has been a classic linkage problem. Until recently, Washington's position was to reject working with Tehran on any subject until Tehran abandoned its uranium enrichment program. Bush subsequently softened his position by linking future direct talks to Tehran's agreeing to suspend rather than cease enrichment. But in the meantime, the United States paid a price in Afghanistan. At the beginning of the war there, Tehran had been helping quietly to neutralize the Taliban on its Afghan border. Bush's linkage prevented such cooperation thereafter. For several years, Bush also refused to talk to Tehran about Iraq, though Tehran was a part of the problem there and had to be part of the solution. Since then, there have been talks, but at a low level. By 2008, it began to look as though Bush's policy of linkage had given more leverage to Tehran than to Washington.

Linkage has a mixed to negative history when based on policy considerations alone, but it actually provides somewhat better leverage when based on politics. If a target state realizes that American political realities simply won't tolerate aid without reforms, for example, it is more likely to be flexible on reforms. Politics is a fact of life that others can't dismiss the way they might a policy. In general, however, linkage tends to delay compromises all around.

Linkage should not be confused with a decision by both sides to put all issues up for discussion. This allows both more items to trade and can thus facilitate settlements. Leaders incline toward this approach when they have a plateful of consequential differences and desire to move beyond deadlock. During the Nixon years, this happened to a certain extent between the United States and the Soviet Union. Today, some advocate the same all-cards-on-the-table approach between Washington and Tehran.

Beneath the dilemmas discussed above is a deeper political and policy dilemma: whether to play an issue for all it's worth politically or act in the national interest regardless of political consequences. Obviously, presidents always insist they're acting in the national interest. No memoir by any senior official will ever confirm that a president subordinated the national interest to his own narrow political ambitions. In fact, the memoirs rarely address politics, though it is the silent and often controlling presence in every foreign policy debate, public and private. That's because even contemplating such a trade-off is the ultimate political sin, if the person contemplating it is caught.

Vietnam and Iraq are the best and most controversial examples.

Of course, presidents Johnson and George W. Bush believed in their causes, but they also played politics once it became clear that they were far more likely to lose than win these wars. At that point, it's fair to argue, both persisted less for their avowed reasons of national interest and more to ensure that they wouldn't personally be held accountable in history for losing. Both understood after several years of war that they were playing long shots by staying the course, but preferred the risks and costs of not losing to the political certainty that defeat would destroy their legacies. The ultimate domino was not Southeast Asia in Vietnam or the Gulf in Iraq, but the White House itself. Johnson and Bush both decided to do whatever was necessary not to lose and to pass the tar baby on to their successors.

On the other side, there are comforting cases of presidents resolving tough foreign issues, though they knew they would take political lumps. The Panama Canal Treaty of 1977 enhanced the U.S. position throughout Latin America, but critics in Washington battered Jimmy Carter for abandoning American sovereignty there. George H. W. Bush worked hand in glove with Gorbachev to dismantle the Soviet empire against a constant barrage from conservatives in his own party as well as from liberal Democrats, which damaged him in the 1992 election against Clinton. After more than three ugly years of stalling and making excuses, Bill Clinton finally did take strong military action in Bosnia through NATO, and did so in the face of public and congressional opposition.

Two recent examples, however, show how readily presidents may now forsake good economic and foreign policy under political pressure. The first was Bush's decision not to clear the path in 2005 for the China National Offshore Oil Corporation's takeover of the American oil company Unocal. This proposed deal was fully consistent with Bush's economic policies and his commitment to long-term economic ties with Beijing. It posed no threat to our energy or national security interests. But Congress erupted, and Bush, seeing legislative defeat ahead, quietly sat still while Beijing withdrew the bid. He did the same again the next year when a Dubai company tried to purchase the management rights to a number of American ports. Congress and especially some Democrats went crazy and insisted that the deal would undermine American security. Bush did retort that security for the ports would remain under American control, but he essentially backed away from a fight he probably wouldn't have won anyway.

Perhaps the most damning example of politics severely damaging the national interest has been America's continuing dependence on Middle Eastern oil. All agreed that this dependency made no sense in terms of national security. It required tying America's fate to fundamentally fragile Arab regimes. Since the first Arab oil embargo of 1973, the most sensible alternative would have been to raise gasoline taxes, squeeze auto manufacturers to manufacture more fuel-efficient vehicles, and develop energy substitutes. But because these actions would require taxpayers to dig more deeply into their pockets in the short and medium term, no president would stake his presidency on what needed to be done.

To manage this problem is difficult because of the taboo on discussing politics openly. Politics, then, gets swept under the rug and triumphs in silence. Thus there's no chance to convince presidents that good policy is good politics, that Americans will respect and vote for a president who subordinates his political future to the national interest.

On the verge or in the midst of crises, leaders also have to choose between quick capitulation or compromise before wagering much power and prestige or betting the whole ranch on unbending firmness. The right choice has generally turned on making one judgment correctly: which side's vital or crucial political interests were unarguably at stake; or, to put it another way, which side had more to lose than the other. If one state's interests dictated doing whatever was necessary to prevail, the other was well advised to move on.

On April 1, 2001, a U.S. EP-3 reconnaissance plane on routine patrol flying in what the United States clearly considered international airspace collided with one of two Chinese F-8 fighter planes dispatched to intercept it, killing a Chinese pilot in the process. The U.S. plane was forced to land on Chinese territory. Bush rejected the advice of his hard-liners to threaten the most serious consequences if the crew were not released immediately. China demanded that the United States "bear full responsibility" for the incident and apologize.

At first, Colin Powell insisted that the United States had "nothing to apologize for," even though he and Bush expressed regrets for the death of the Chinese pilot. Then, Bush wrote a letter to the pilot's wife, and finally, on April 11, the United States sent a letter to China saying it was "very sorry" for the pilot's death, and "very sorry" that the U.S. plane hadn't received oral clearance to enter Chinese airspace when it was forced to land. After a little under two weeks of growing tension, China released the crew; much later, it returned the plane, disassembled.

Without the "very sorrys," the situation probably would have escalated far out of proportion to its inherent importance. To be sure, the Chinese acted outrageously. But Beijing had all the cards to play short of major confrontation. The only path for Washington to win would have been escalation, and the issue wasn't worth it. Predictably, the incident has been almost totally forgotten.

On March 23, 2007, Iran captured and detained fifteen British Royal Navy sailors and Royal Marines on patrol in the Gulf, claiming they had strayed into Iranian waters. Tehran called the presumed British intrusion "blatant aggression." The British prime minister, Tony Blair, warned that failure to release the crew would require the United Kingdom to move into a "different phase." On March 28, the United Kingdom froze all bilateral business with Iran. Iran demanded that the British concede they were at fault as a condition for releasing the crew. On April 2, Tehran stated that all the captives had confessed to an illegal intrusion into Iran's territorial waters. Two days later the Iranian president, Ahmadinejad, announced that Iran had "pardoned" and would release the detainees, and said this was in response to a British letter stating that such an incident "will not happen again." London did this one quite slickly by not directly apologizing itself.

The Cuban missile crisis of 1962 was, of course, the quintessential example of when not to fold and simply finesse a problem, and how to manage the most dangerous test of wills. The Soviets never would have deployed their missiles secretly to Cuba unless they thought the United States under Kennedy to be weak-willed and unlikely to respond. Kennedy had to be tough and make U.S. vital interests clear, and he did. But at the same time as he took the world to the brink of nuclear war, he offered concessions to let the Russians save face, such as his pledge not to invade Cuba and his later withdrawal of missiles from Turkey.

Even in the model case, even after the dilemmas are sorted out and all other pieces of the puzzle are thoroughly examined, there is a kind of weariness to the foreign policy business. The very big problems often are truly intractable. Maybe Iraq will lapse into civil war after we mostly depart, or maybe things will resolve themselves so that Iraq doesn't become a bother or a temptation to its neighbors. Perhaps the United States and NATO eventually will discover they need a face-saving exit from the brutal history of Afghanistan. And it might turn out that Iran will develop peaceful nuclear capability under international inspection and stop short of manufacturing nuclear weapons.

The uncertainties of foreign policy are indeed wearying. Even with the best of American policies and with the optimal cooperation of other key states, our presidents and our leaders can't really know with confidence how hellish situations like Iraq, Iran, Afghanistan, and Pakistan will evolve. They will fret over their uncertainties and doubts in private, but be reluctant to share them publicly for fear of being tagged as defeatist. Their critics will bombard them throughout, demanding clear answers and success.

My foreign policy colleagues and I add to the difficulties by continuing to talk as if the problems can all be resolved and failures avoided if presidents only did this or that. It's very hard somehow even for nonpoliticians to tell Americans the truth—that our presidents can do what they can with American policy and power, and they can do it with more or less skill, but that final results often depend more on others, especially on those we are trying to help or hinder. Presidents will not be able to escape this bleakness, but they can give themselves better chances of success and cushion the ride.

FOREMOST IS FOR PRESIDENTS to realize that foreign policy is their most precious instrument for defining and handling problems and opportunities. They should not treat it as a gimmick or a political speech to make themselves look good for a week and deflect political adversaries. Policy is their chance to integrate intelligence, power, and strategy—means and ends. It's the main design for ensuring that power is sufficient unto ends, and that goals are thus attainable. It is the principal method of gathering domestic support, clearing paths abroad for power, and enabling power to spring into action.

Policy should be thought of as a galvanizing mechanism, not as the power to dictate or the right to lecture. Galvanizing power is the means by which to trigger coalition-building and action in the desired direction. It most certainly doesn't absolve presidents of the necessity to compromise in order to forge needed coalitions. The trick is to transform necessary compromises into opportunities for leadership and problem solving.

After the coalitions are in place, the coalitions and the policies that guide them have to be thought of as giving the United States leaning power, much more than threatening power. And it must be realized as well that leaning power takes patience, and that policy must help buy the required time for patient leaning.

Leaning power can work. It's based on America's superior resources and its advantageous strategic positions around the world (others still need us against bigger threats). In time, foreign leaders will feel the weight of what America and its coalition partners can do to help or hurt them. In the last analysis, however, the effectiveness of the leaning will turn mainly on those being leaned upon, and neither presidents nor the American public should forget that.

America's leaning power will be significantly diminished if presidents do their typical thing: try to ride two opposing tigers in tough situations and try to have it both ways. Policy dilemmas have to be resolved, one tiger or the other has to be chosen, or U.S. power will be set at cross-purposes and cancel itself out.

All these measures will increase the odds that American power will be effective—but some problems can't be fixed even by the most cleverly devised policies. Too often, presidents have reacted to these problems from hell by throwing good lives and money after bad and simply passing the hell on to their successors. Too often, they resist small and manageable setbacks, make deeper commitments, and create bigger, more costly, and far more dangerous situations. Wise policy can help manage and explain the setbacks, and provide new time to adjust and perhaps eventually to prevail.

In the end, American power and diplomacy can be only as good as the foreign policy on which they are based.

CHAPTER 12

U.S. POLICY AND POWER
IN THE MIDDLE EAST

Just as Asians sought U.S. protection from China after Vietnam, Arab leaders now need American power to check Iran in the wake of the war in Iraq. This is your opening to turn the Mideast around, and you have the power to do it. All you need is the right strategy, and it's staring you in the face.

With the United States exiting Vietnam in the early 1970s amid wails the world over for the passing of the American era, President Nixon and his policy soul mate Henry Kissinger accelerated two precariously competing policy lines. At home, they hyped their dire warnings: Defeat in Vietnam would trigger the fall of friendly countries or "dominoes" to communism and the decline of American credibility everywhere. Overseas, they ramped up a diplomatic campaign to prevent the collapse of those dominoes. Move one was designed to blame the looming defeat in Vietnam on critics of the war. Move two was to restore U.S. power worldwide—and they did just that.

Nixon and Kissinger's game plan sprang from three central strategic facts: As the United States began its withdrawal from Vietnam, most Asian leaders saw the Americans' defeat as a Chinese victory, and China as a growing threat to their security; Asian countries still believed the United States was the only possible counterweight to

China, and a relatively benign one; and therefore, whatever problems Asians had with Washington, they needed a strong America in their neighborhood. They feared the Vietnam debacle would prompt an American retreat from the region.

Nixon and Kissinger manufactured streams of diplomacy that went beyond Asia itself—a cascading display of America's unique role and power in the world. In Asia, they strengthened ties with key nations such as Japan and South Korea, and generally promoted new military aid and training programs. In the Middle East, they maneuvered for and rapidly constructed a cease-fire pact between Israel and Egypt, which started these bitter rivals down the road to a cold yet stable peace. And most dramatically, they established big-power triangular diplomacy with the United States as the pivot between the Soviet Union and China.

Almost all of Asia welcomed and marveled at these fireworks, which they realized could be performed only by the United States. Just a few years after that dark day when U.S. helicopters lifted the last Americans off its embassy's rooftop in Saigon, after that unforgettable scene of defeat, the United States had already restored its position in Asia and was stronger than at any time since the end of World War II. Nixon and Kissinger helped create a new Asia by buffering the last Cold War shocks and reducing the sense of regional danger. They thereby laid the security foundation for the dramatic period of Asian stability and prosperity that was soon to arrive. And they helped transform China from Asian enemy number one into a competitor and sometimes a partner of the United States.

No one predicts such cosmic transformations for the Middle East in the twenty-first century. Arab leaders today are more ambivalent and uneasy about their dependence on America than were the Asians of the 1970s, and far more ambivalent about American power. There's a degree of endemic anti-Americanism, exacerbated greatly by United States' deep ties to Israel. Also, these Arab states lack the political discipline and stability of Asian countries. Their politics churn with ancient hatreds and their rulers sit uneasily on their thrones. Arab and Muslim political cultures are deeply torn between needing Western protection and resenting Western influence.

Nonetheless, the United States has significant strategic opportunities in the Middle East today, and they do resemble the ones open to Nixon after Vietnam—contrary to the prevalent pessimism of regional experts. Sunni Arab rulers today would tremble to see the United States defeated in Iraq and the rise of a Shiite Iran, every bit as much as Asians worried about the emergence of China. A weakened America leaves them precariously exposed to extremism, subversion, and insurgency. They don't want America to be diminished, especially as measured against an ascendant Iran. Most Arab governments see the United States as the one nation capable of counterbalancing Iran and related threats, just as Asians viewed America as the sole counterweight to China. Most Arab leaders long for an American policy in the region that they can embrace and that promises success—but such a policy is precisely what they don't have.

THUS ARAB LEADERS FEEL themselves threatened from without by Iran and from within by extremists as well as moderates. They see the United States as the only nation that might be able to help. That is America's strategic opportunity.

And Washington has more than just the opportunity; it has the power. Experts on the Middle East tend to devalue this power because they now significantly overvalue Iran's and often misunderstand America's.

First, these experts think American power is not equal either to the magnitude of the problems in Arab societies or to overcoming the natural resistance to outsiders generally. They are correct on those counts. Washington hasn't got the power all by itself to solve the problems facing Arab leaders. They'll sink or swim mostly by what they themselves do. But Washington has significant power to help deal with these problems, and in any event has its own interests in the region and must try as best it can to patch together a coherent policy to protect those interests.

Second, experts underestimate both America's strengths and the weaknesses of countries in the region. Specifically, they see America declining and Iran rising. They fail to appreciate that Iran, unlike China, is a third-tier country—still poor despite its oil, and backward militarily and divided politically. Its economy rivals that of Texas but with more oil and twice the unemployment rate, while its conventional military capability barely compares with the U.S. National Guard and reserves. Its power in the region has grown in recent years for one reason above all—the blunders of the Bush administration, which Tehran has exploited rather competently. Tehran also profits unduly from the nonsensical Washington rhetoric that portrays Iran as the region's great threat and future powerhouse. As they did during the Cold War with the Soviet Union, Washington's politicians and policy experts simply love to lavish undeserved and free power on our enemy.

Although Washington is not as strong as it once was in the region and can't issue orders to anyone, it nonetheless remains the paramount power there. No other state or group of states approximates American power. The European Union equals America's economic clout in the area, but possesses almost no military punch, and lacks the unity of purpose and America's standing with all regional parties. The EU can't play a leading diplomatic role. Regional rulers treat it as a kind of footnote to Washington or a minor safety valve in disagreements with the United States. China counts only as a country with a veto in the UN Security Council and as an oil customer. It's not a factor for the foreseeable future, except with regard to Iran. Russia has lost its Cold War role in the Middle East as the power second to the United States; but it can still be very helpful or harmful, especially with regard to Iran. It remains Tehran's main supplier in nuclear matters. To Arabs and Iranians, America is still number one—for all the limitations on its power.

THUS THE UNITED STATES has the opportunity and the power to reestablish itself in the Middle East; what it lacks is a strategy and a policy to exploit its opportunity and power.

American goals in the region should be modest, but not too modest, as befits the region's complexity and U.S. interests. There are three major and inescapable goals. The first is to secure the oil supplies for at least another decade, while Washington presumably reduces its dependence on them. The second is to contain and neutralize international terrorists who thrive in these parts. The third is to help Israel ensure its security.

President Obama must factor in the limits of American power. Perhaps the key restraint is the anxiety of Arab leaders over the fragility of their power within their own countries and their concomitant fear that overt cooperation with Washington will only increase their vulnerability at home. The Saud family bribes its opposition with money and quiet support for religious extremism. President Mubarak, like the pharaohs, concedes little to adversaries and governs with relentless force and toughness. King Abdullah II of Jordan doles out dollops of power to rivals, but draws lines that can't be crossed.

While experts argue about exactly how vulnerable these regimes really are, the fact is that their rulers won't gamble with their hold on power. None can be counted on to make the internal reforms we urge upon them, and they resent our even asking. Nor are they readily inclined to make compromises in negotiating, especially those perceived as benefiting Iran's Shiite leaders or Israelis. Thus America's friends in the region speak with forked tongues and act with annoying and duplicitous caution—not totally unlike America's policy.

To counter Arabs' doubts and fears, Washington needs a policy that will give the appearance and the reality of the wind being at America's back again. Countries in the region will have to have the confidence that Washington's new approach will work and make them more secure. This strategic momentum will emerge from

America's basic power source in the region, which is about the same as America's power source worldwide: the common knowledge that the United States, far more than any other country, can forge the necessary power coalitions to cure or mitigate problems. If Iran wants security and economic development, the road runs through Washington. If Tehran threatens its neighbors, American presidents can deal with that as well, through economic sanctions and stronger measures. If Iraqis are determined to just kill each other, there's nothing Washington or anyone else can do; but if the warring parties want peace, Washington alone can assist. If Palestinians and Israelis choose peace, only the United States can broker and guarantee it. If Syrian leaders yearn for respectability and to solidify their power at home, Washington can either grease the wheels or apply the brakes.

NEXT COMES THE EVEN harder portion—reinforcing this overall leadership momentum with plausible policies for the specific major issues in the region. During the Bush years, most regional leaders had little idea what Washington was up to, or if it even knew what was going on.

President Obama does have plausible—not great, but plausible—policies to try. On Iraq, he has the Bosnian model to resolve differences among the various parties by decentralizing power. On Iran, he can use the Libyan model, whereby Washington and Tripoli put all cards on the table and traded them most satisfactorily. For the Israeli-Palestinian talks, there is useful precedent in Northern Ireland, where time was taken to nurture political constituencies for compromise. There's no model for Syria, only a sense that the ruling Alawites really want American help, but slowly and without much public embrace.

WASHINGTON, THEN, HAS TO decide how to prioritize and relate these individual efforts. Specifically, the president has to choose which door to open first both in terms of American needs and in terms of likelihood of success in creating diplomatic momentum.

Overall talks with Tehran must get started very soon to show that America is back and unafraid, but years will pass before satisfactory agreements can be achieved. As for Syria, Washington can let it dangle for a while so long as Damascus sees America on the move again. Nor is it wise at the outset to press Palestinians and Israelis for concessions that neither side is prepared to offer at this time. What is essential is to initiate a plausible diplomatic process immediately. Iraq has to be the first door to pry open.

EVEN WITH THE SITUATION in Iraq having quieted, America's central role in this unhappy war drags down everything else. With huge American resources, prestige, and power still engaged there, and with the president's time and political capital still claimed by Iraq, the first steps must be to end all major American combat operations, step up internal political talks among Iraqis, and leave in a way that gives the Iraqis a decent chance for stability.

Even with the war thankfully quieting down, the president still has to spend enormous time and power on Iraq. Americans are still fighting and dying there. Direct annual costs now exceed $100 billion. The U.S. forces elsewhere still can't mount serious military operations because of the requirements in Iraq. Americans still demand, though more softly with casualties low, an end to this six-year struggle—presumably, as long as it doesn't conclude badly.

Upon taking office, Obama quickly reinforced the withdrawal timetable agreed on by Bush and Maliki. This agreement called for a phased withdrawal of U.S. troops from combat, from combat regions, and then a complete departure of American forces by the end of 2011. Even as the U.S. withdrawal is underway, Obama has to take special care to prepare Americans for difficulties in Iraq and to stress that results are now up to the Iraqis themselves. He hasn't done enough to foster an overall political deal among Iraqis or to engage Iraqi leaders in commitments to oppose terrorism from within their borders.

As American troops reduce their presence and involvement, Iraq's built-in tribal and religious tensions come once again to the forefront. The Kurds find themselves at odds with both Shiites and Sunnis over oil rights and power. Terrorist acts mount between Sunnis and Shiites. The Obama team seems to be doing little either to use remaining U.S. leverage to push for a political settlement or to prepare Americans for the Iraqi turmoil ahead.

A political settlement is, of course, the key to this situation. It is necessary to create conditions for security and economic measures to take hold. Otherwise, the risks of a new civil war will stop all progress. The working model for this can be Bosnia.

As president of Yugoslavia, Tito contained the historical hatreds among his Croatians, Bosnian Muslims, and Serbs and the larger Yugoslav Federation by giving slices of the power pie to all and retaining all military and police power in his own hands. After Tito died, Slovenia and Croatia declared their independence from Yugoslavia and established their own separate states. Thereupon the Serbs, with the overwhelmingly dominant army, began killing Muslims and Croats and ethnically cleansing Muslim and Croat areas in Bosnia in an effort to expand Serb sovereignty. The slaughter went on for several years until the victims were armed and the United States and NATO intervened.

President Clinton and his envoy Richard Holbrooke pushed the adversaries into negotiations near Dayton, Ohio. They tried, initially, for shared powers in a workable central government. But there was no real chance of making Sarajevo, the capital of Bosnia and Herzegovina, into a strong central government, given the history of ethnic, religious, and civil war and recent mass murders. Thus Holbrooke turned to a federal solution.

The federal idea was to divide constitutional powers in a united country along geographic lines, between a central government with

limited powers and strong regional governments that largely controlled their own affairs. Holbrooke pressed for a united Bosnia
with two federations, one for the Serbs and a combined one for the
Muslims and Croats. The underlying reality was a tripartite government with the groups retaining their own armies as the only reliable means of self-protection. The war and the Serb policy of ethnic
cleansing had already driven many from their mixed neighborhoods
and cities into separate enclaves. The three groups signed on to the
federal approach with two pivotal reinforcements: The European
Union promised economic aid if the parties behaved; and a sizable
contingent of U.S. and other forces was dispatched with the firepower and authority to police a peace. This approach has held to
this day.

A strong central government cannot work for Iraq at this time
any more than it could in Bosnia. Centralized power in Iraq would
mean only one thing: that the Shiite majority would have dominant
power, which the Sunni Arabs and Kurds would never accept peace-
fully. There isn't sufficient trust among the three groups for this
form of government, and besides, the present central government is
mostly nonfunctional, corrupt, and in Shiite pockets.

The only form of government that stands a chance of keeping
Iraq a united country is a federal system. Obviously, Iraqis would
have to decide on the exact form and details themselves, but at its
core, federalism in Iraq would mean a relatively weak central government responsible for border defense, foreign affairs, distribut-
ing oil revenues on constitutionally agreed terms, and oversight
of currency and banking. Real legislative, administrative, judicial,
and internal security powers would reside within the regions, be
they three or more. This would be particularly gratifying to the
Kurds, who already have and would continue to retain their own
regional government. Increasingly, Sunnis seem to realize that even
a reduced piece of territory or region under their control would be
far preferable to being a permanent and often abused minority in a
Baghdad government run by the Shiites. The Shiites also increasingly appear to understand that while federalism would deny their

wish for complete control over all Iraq, they wouldn't get that control anyway without a constant fight, and that a continuing civil war would put their sizable oil revenues in jeopardy.

As in Bosnia, while the final decisions would be up to the Iraqis, the United States would be more than a mere mediator; it would have to press and cajole and remind all that some U.S. troops would stay to help keep the peace only if the fighting died down reliably as a result of a political settlement.

As for Iraq's neighbors, they wouldn't be thrilled with federalism. They regard it as a model for breaking up their own states. But if the Iraqis themselves bought it, their neighbors would go along. And it's important to remember that if neighbors' tolerance for Iraqi federalism spelled the difference between success and failure for the United States in the Gulf, most neighbors would not want Washington to fail. President Obama, then, could lock in this support by sanctifying Iraq's political settlement with a regional peace conference.

This new strategy would also make it easier for Washington to assemble the requisite power coalition to reassure Iraqis that they will have the necessary technical aid and international support. The coalition would not be a cumbersome Tower of Babel. The key participants would be the United Kingdom, France, and Germany for the European Union, and Saudi Arabia.

If Iraqis nonetheless rejected this federal model, and if internal wars continued, President Obama could and should move ahead with ongoing U.S. troop withdrawals, putting the responsibility squarely on Iraqi shoulders. Most Americans would be sympathetic to this careful withdrawal process. Most would feel that six years of war under Bush and however many additional years needed to complete a withdrawal would be enough—and that the time had come to pay more heed to other critical issues.

The South Vietnamese couldn't put together a government that its huge army was willing to fight and die for against the smaller and far more motivated North Vietnamese. The South Vietnamese have suffered the consequences to this day. Iraqi leaders have to realize that if they can't come to terms with one another and protect their own people, their fate would be worse than that of the South Vietnamese. The Vietnamese live without civil strife and with economic gains under a repressive communist dictatorship. The Iraqis could sink into a new sea of civil war and perhaps regional war as well. And for all the grief this would cause to U.S. interests, Americans would recover from "defeat" in Iraq far more quickly than the Iraqis, just as Americans rebounded faster than did the Vietnamese.

THE STRATEGIC MODEL FOR approaching Iran is Libya. If anything, we regarded its leader, Colonel Muammar Qaddafi, as worse than the hawkish Iranian mullahs and President Ahmadinejad. The colonel had chemical weapons and was closer to nuclear weapons capability than Iran is today. He was a major supporter of terrorism, providing money, training, and safe havens for attacks against Americans and others. He was fostering insurgencies and wars in Africa. He, too, demanded the destruction of Israel. Every indictment we hurl at Iran today we once charged against Libya. Our leaders proclaimed almost unanimously that they could never reconcile with these Libyan monsters. Yet, in 2003, President Bush concluded the most significant agreement of his administration with none other than the untouchable colonel. With almost no comment, the United States did not block Libya from joining the UN Security Council in 2008 and eventually ascending to its presidency.

Not surprisingly to diplomats, each side made concessions, and the arrangement met the most critical demands of both sides. Qaddafi stopped Tripoli's extensive involvement in terrorism (insofar as U.S. intelligence could tell), provided information on terrorists to Washington after 9/11, took responsibility for blowing up Pan Am Flight 103 over Lockerbie in Scotland, compensated the families of the dead, renounced weapons of mass destruction and destroyed related facilities, opened everything up for full inspection, provided information on other nations involved in these illegal WMD programs, and generally uttered a variety of soothing incantations. In return and after further progress, the United States eased various economic sanctions and moved to normalize relations. Bush made no demands that Qaddafi tenderize his nasty dictatorship. Bush, the self-proclaimed crusader for democracy in the Middle East, simply agreed to leave the colonel's total and oppressive political power intact.

The deal was simple: Qaddafi met all of Bush's security demands, and Bush agreed to Qaddafi's staying in power and helped to strengthen his regime economically and diplomatically. Without doubt, it was the best piece of diplomacy of the Bush administration.

Inevitably, experts contend over why Qaddafi did this, but there's little hard evidence on his motives. Maybe he worried about U.S. forces attacking his country as they had just done in Iraq. But the fact is that he had started down this peace track before the invasion of Iraq, and according to intelligence sources, had terminated his aid to terrorists in 1993. Maybe Qaddafi's oil economy was suffering badly from the economic sanctions. Indeed, it was not in good shape, but it wasn't in bad shape either; oil revenues were mostly holding up. Business friends of Libya maintain that Qaddafi simply changed his mind and wanted a better and more peaceful future for Libya. But the truth is we don't know, and we won't be able to crawl into the minds of Iran's rulers, either—or know just how secure or insecure they feel.

We do know that Qaddafi began signaling a change of course long before the deal was actually concluded. Experts insist that such an opening has not been forthcoming from Iran. But they are wrong. Almost unnoticed, Iranian leaders began helping the United States at the outset of the Afghan War with border control and other matters. And most interestingly, only weeks after the fall of Baghdad, an Iranian source offered Washington what looked on the surface to be an incredibly far-reaching grand bargain—much like Qaddafi's opening gambit. In its offer, transmitted through and perhaps modified by a Swiss diplomat, Iran undertook to tackle all major American concerns, including "full transparency for security that there are no Iranian endeavors to develop or possess WMDs," "decisive action" against terrorists, coordination of efforts in Iraq, ending "material support" for Palestinian militias, and buying into the Saudi proposal for a two-state peace between Israel and the Palestinians. In return, the Iranian document called upon the United States to recognize Iran's "legitimate security interests," end economic sanctions, and provide access to peaceful nuclear technology.

Richard Armitage, the number two at the State Department at that time, told *Newsweek* in 2007 that it appeared to the administration that the Iranians "were trying to put too much on the table" for serious negotiations to occur. But this explanation is hard to follow. Why would this opening be a ploy and too big a package to negotiate, whereas Qaddafi's similarly wide-ranging offer was not? Perhaps the Bush team believed the memo came from an unauthorized peace faction in Tehran, not the bosses. The problem was that the team made no effort to find out.

Bush could have probed the Iranian memo without removing his military options. There was and is value in Iranian hotheads gazing into nearby waters and seeing this: U.S. carrier strike groups with aircraft and hundreds of long-range cruise missiles capable of destroying every single military and economic target of value in Iran—without one American soldier setting foot on Iranian soil. Nor was there need for Bush to lift economic sanctions at this point. The time was ripe then and is even riper now to wheel in diplomacy. In fact, in mid-2008 the Bush administration let out the word that it was considering opening a U.S. interests section, or a minimal diplomatic presence, in Tehran, the first direct American presence in that country since the 1979 hostage crisis.

From the outset, President Obama pushed hard rhetorically to get talks with Iran started. But neither side budged on substance. More portentously, the split between hard-line clerics and the more moderate middle class in Iranian politics exploded publicly in the highly disputed 2009 Iranian presidential elections. In the short run, the behavior of the hard-liners in power would complicate the prospect of talks with Washington; in the long run, the public display of the split in Iran would open the way to more power for moderates who wanted to talk with the United States.

Meantime, Obama has been slow to make two key strategic decisions: One is on the scope of the negotiations—whether everything is on the table or just one or two issues for each side. The other is whether there will be any flexibility regarding Iran's uranium enrichment program. On the first question, the answer is the more the better. If each side can't talk about what it wants, it will spend all its time obsessing about that and not negotiating. Also, the more items on the agenda, the more chances for trade-offs. Finally, the fact is that there are a lot of irritants on both sides. Providing responses to all irritants helps clear the air.

The nuclear question is explosive. Washington doesn't want Iran to have any enrichment program whatsoever, for fear of Tehran's cheating and hiding enough weapons-grade material for bombs. And perhaps if the overall settlement is good, Tehran might relent on this—but it is not likely. Qaddafi may have felt threatened by Israel, but Iran feels the heat from several nuclear neighbors. It worries about Pakistan (because of the Taliban) and Israel for starters, and Iranians still fret about the Iraqis for historical reasons as well. (Several Iraqi leaders told me last year that they would favor Iraq's having a uranium enrichment program just in case, and regardless of what happens with Iran at this juncture.)

Tehran will maintain, and correctly, that as a signatory of the Nuclear Nonproliferation Treaty (NPT), it has a right to an enrichment program as long as it allows UN inspections. Also, Tehran will argue that Washington has already abandoned its strict adherence to the NPT in recent deals with India, Pakistan, and North Korea, and the Iranians would be right about that, too. It's highly unlikely that under these circumstances, Tehran will abandon its nuclear fuel program simply in return for Washington's helping it back into the international economic community. Nor will it buy the argument heard in Washington: "You Iranians are bad guys, and the Indians and Pakistanis are good democratic guys." At some point, then, President Obama will have to choose: either require no uranium enrichment or allow enrichment under the strictest of safeguards and inspections, including sending all spent fuel to Russia.

Obviously, President Obama should push for a ban and not show flexibility at the outset, but he may feel he has to soften somewhat at some point, given America's acceptance of Pakistani and Israeli nuclear weapons. He may also want to trade for concessions on other important issues, including the following: ceasing support for insurgents in Iraq and supporting a political agreement there, helping against the Taliban in Afghanistan, stopping aid to all terrorists such as Hamas and Hezbollah, and at least being neutral on Palestinian-Israeli peace. It's impossible to shove Tehran completely out of Iraq, because the principal power there is now—and will continue to be—the coreligionist Shiites. Also, most Iraqis would want an Iranian buy-in to stabilize any Iraqi settlement.

In return, Washington will have to remove economic sanctions, open the international economy to Iran, and pledge noninterference in Iran's internal affairs. The economic card plus diplomatic acceptability turned out to be crucial for Qaddafi. But they were insufficient to move Qaddafi toward political tolerance—and this arena is likely to be off bounds for Tehran as well. Americans should maneuver for medium- and long-term movement on this front.

Washington could use diplomatic help in these efforts. The fundamental partners would be key European Union countries and Russia, a major arms and nuclear supplier to Iran. From time to time and on particular issues, it would be helpful to include India, Saudi Arabia, Japan, and China, all important traders with Iran.

The main task for this diplomatic coalition is to get Iranian leaders back to thinking along the lines of the grand bargain some of them seemingly proffered in 2003. The United States has the strength both to make such bargains and to respond effectively to disappointments.

. . . . .

DAMASCUS POSES NOTHING BUT puzzles for U.S. strategists. Syrian leaders, like China's, are fixated on retaining internal power, but unlike the Chinese, they're not nearly as preoccupied with participating in the global economy or spurring domestic economic growth. Therefore, Syrian leaders do not feel compelled to establish good relations with the United States simply because it's the world's economic leader. Syria makes few concessions in its foreign or domestic policies in order to gain the good opinion of humankind. Its people live very modestly, and the reigning Assad family seems quite content with slow and modest growth. So far at least, the Assads have given up nothing on the political side to buy speedier growth, and this strategy has not appeared to lessen their stranglehold on Syrian politics. The Assads and the Alawites do not think like the Chinese Communist Party.

Nor do Syrian bosses deal with their domestic situation like the Iranians. In Iran, the ayatollahs, mullahs, and Revolutionary Guards have to bully their people all the time, and the people nonetheless continue to openly defy them. In Syria, however, President Bashar al-Assad and the former leader, his father, the late Hafiz al-Assad, have so institutionalized and ingrained spying and security that there is little protest or resistance. This relative quiet is all the more remarkable given that the Assads come from the tiny, ruling Alawite minority. This is a small cult, neither Sunni nor quite Shiite, whose historical base was the northwestern mountains of Syria. The Alawites took control of the country in 1970 and have dominated its huge majority Sunni population ever since. The Alawites confronted one major rebellion in Hamah in 1982, put it down by killing almost everyone in sight, and then seeded the town's soil with salt. They haven't had a front-page problem since. That is not to say that the Alawites sleep well at night.

It's hard for Washington to get a power handle on Damascus. The United States can't threaten credibly to promote political opposition there; the Assads have the place locked up. The Israelis, with or without American help, could destroy the entire Syrian army, but the Assads know that and don't do anything to provoke Israel. They even take some care not to be too blatant in letting arms and men slip over their border into jihadi camps in Iraq, lest the United States respond by hitting their border posts.

Thus the problem for Washington in trying to establish its power in Damascus is that in Syria—unlike most wannabe countries—heads aren't turned by most trade and investment lures. In fact, the Assads seem leery of broad and swift economic development. Such growth would benefit disproportionately the predominantly Sunni commercial classes, giving them new wealth that would inevitably presage demands for new political power. All this diminishes Washington's economic carrots and sticks, which have proved so potent with most other countries.

If Washington can't use an economic strategy as its central wedge, another approach is to work with Israel on the return of the Golan Heights to Damascus in exchange for normalized diplomatic relations and peace, and then to leverage this into improved Syrian-American relations. The Israelis may consider the Golan expendable in the missile age, and the Assads would derive great prestige from its restoration. The quid for Israel would be an end to Syria's support for Hamas, Hezbollah, and other extremists bent on destroying Israel.

If Washington can put any hooks into Damascus, it would be to provide the Assad clan with more internal security, prestige, pride of international standing, and respectability. Whereas other Middle Eastern countries fear association with the United States, Americans can actually strengthen Assad. Although the Syrians lack the elegance and wealth of the Saudis and the culture and position of the Egyptians, their pride and their drive for respectability are fierce. There is something here for Washington to massage. Indeed, the Obama administration has been moving very cautiously in this direction by sending a U.S. ambassador to Damascus to reestablish full diplomatic ties and by selectively lifting sanctions against Syria to permit certain civilian sales.

. . . .

THE PALESTINIAN-ISRAELI NIGHTMARE DOESN'T have to be solved next week, but a serious peace process cannot be set aside. Stalemate there poisons everything else. Repeated failures on this front also diminish American power in the region generally. The diplomatic dynamic of the last decades has been, as they say in diplomatic parlance, counterproductive. On the one hand, Arab and European leaders keep pressuring Washington to drag the two adversaries to the negotiating table, and it eventually does, knowing full well the negotiations will fail—as they do, again and again. On the other hand, and to avoid the failure of the first path, Washington often sits back and does little or nothing, which also inflames the region and suggests American weakness. The Obama administration's answer to this historic dilemma has been to promote a peace agreement between Israel and major Arab states such as Saudi Arabia and to use these as leverage to prompt an Israeli-Palestinian peace agreement. But there's no good evidence that the administration can reach such a broad regional accord or that it would change Israeli-Palestinian politics.

The challenge is to gear U.S. diplomacy to create the right time for a breakthrough. The first move, alas, still has to be herding the ancient enemies to the formal negotiating table. But Washington also needs a strategy to help both sides do two things: compile the trust essential for peace, and help Arab leaders in particular structure the political support essential to make the compromises necessary for peace. The idea is to do in the political arena what hasn't been and can't be accomplished at the negotiating table: nurture political support for leaders to compromise.

Some variation on how the United Kingdom and the United States mediated the Northern Ireland pact of 2006 between Protestants and Catholics might shed light here. The former senator George Mitchell and the policy expert Richard Haass, two of the U.S. negotiators to Northern Ireland, explained that compromise flowed from both sides' being able to "hold onto their dreams" by leaving open Northern Ireland's future status. The Catholics could still dream of a united Ireland, and the Protestants of remaining with the United Kingdom. Mitchell and Haass also related that it was helpful to include warring factions in the negotiations, however much they slowed progress.

These approaches, however, can't be applied wholesale to the Mideast. The Israelis could never agree to preserving Hamas' dream of exterminating Israel and restoring Palestinian sovereignty there. Hamas similarly refuses to accept Israel. As of now, their dreams generate nothing but negotiating nightmares.

That unfortunately leaves the United States and Israel with only one Palestinian negotiating partner: the weak and corrupt Fatah Party. This sad fact creates yet another American dilemma: Fatah leaders will never sign or implement a treaty with Israel that doesn't have substantial political backing among Palestinians, but at the same time, Fatah leaders are unwilling to expend the capital to develop that support. There is no Palestinian leader like President Anwar Sadat of Egypt, willing and able to shoulder the burdens and personal risks of peace with Israel. Perhaps it is because the others remember his fate.

This is where the experience in Northern Ireland has direct relevance. London and Washington strove hard to foster political support for peace among the warring groups. They nurtured the fledgling movements for a cease-fire among Catholic and Protestant women and businessmen. As the troubles quieted somewhat, they fostered an unprecedented period of economic well-being. Many in Northern Ireland were given their first economic stake in peace. These efforts at political coalition-building and confidence-building never ceased.

Washington can do the same and more for Fatah. It will need cooperation from major Arab states such as Saudi Arabia, from the Europeans, and above all from Israel. They will all have to focus on and assist economic development and political constituencies for peace. Probably, this process should begin with an agreed statement of principles: a two-state solution and a Palestinian capital attached to East Jerusalem.

Yasir Arafat walked away from a most forthcoming Israeli offer in 2000 at Camp David because he had done nothing to lay the groundwork among his people for any kind of far-reaching agreement with Israel, almost no matter how beneficial its terms were for Palestinians. With a public still so hostile to Israel, Arafat probably reasoned that even a great pact would look like a sellout. Some of the American negotiators have said in recent years that the Palestinian people would have accepted the offer had they known about it. But that's precisely the point: They didn't know about it. Arafat had not begun to prepare his public, and his opponents could and did easily mischaracterize the Israeli proposal. Palestinians didn't and still don't realize what a good deal they had. Recognition depends on trust, and that has to be developed through a variety of confidence-building measures, especially concerning mutual security. Prime among these moves would be for leaders on both sides to tell their own people in their own native tongues that peace is necessary and good. If they won't take this step, there isn't much hope.

The way to provide some courage to Palestinian leaders is for Washington to persuade Israelis to reaffirm the Camp David terms and related peace terms publicly. It meets virtually all Palestinian conditions and is obviously very close to what a final agreement must look like. But Israel's quid pro quo has to be a commitment and a plan by Fatah to build constituencies for peace. Washington will also need to fashion a power coalition of key European states plus Egypt, Saudi Arabia, and Jordan. All of them have to buy into most of the strategy to make it work.

In any event and despite the best-laid plans, Fatah may be hopeless, and Washington and Israel may have to move on and deal with Hamas and the terrorists as best they can. But the United States does have leverage with Fatah; America is the party's only lifeline at this point, so Fatah's leaders may just realize this is their last chance before they sink into irrelevance.

President Obama, however, will have to buy the necessary time to set the proper foundation for agreements. Bush helped out here by crossing an important Rubicon and pledging the United States' support for a two-state solution, a Palestinian and a Jewish state. That was crucial for the Palestinians, and the Israelis have not rejected it. Israel's contribution to time-buying should be to show flexibility on expanding existing settlements and building new ones. It is the most neuralgic point to Palestinians. The only real way for the Palestinians to help would be an all-out effort to stop terrorist attacks. Nothing would mean more to Israelis. Initial U.S. diplomacy should focus on these tangible steps.

PRESIDENT OBAMA'S ADDRESS TO the Muslim world in Cairo on June 4, 2009, cleared the air some and showed American sensitivity to Arab feelings and history. But the speech did little to explain how the president would use America's power to help solve the region's problems. It was more about music than about the notes of how the president would get things done.

Here is the game plan I would offer President Obama to kick off an overall Mideast diplomatic surge. It starts with a presidential speech that explains the new American strategy. But it has to be much more than just a smart speech. It has to sky-write that America is back, knows what it's doing, and has a plausible and compelling strategy, one that has support in the region and at home, one that makes the case for time, patience, and realistic expectations. It can be done—with some political savvy and courage.

In that speech, you'll remind all that two of America's most serious and immediate security threats spring from this region—the threats from international terrorists and the threats to our overall economy from our dependence on oil. We can't throw our hands up and walk away from either. Nor should you put America in its usual unrealistic corners—promising early solutions that turn to gossamer or making chest-thumping threats that turn into real wars.

American strategy in the Middle East will hinge on the needs of the countries of the region for peace, stability, and fairly distributed economic growth—and the realization of these nations that the United States alone has the power to help them, to lead, to build coalitions to address these problems. The key moves are twofold: first, to disentangle carefully, but with determination, from the Iraq War; and second, to treat Iran both as the main regional threat and as a potential competitive partner. Washington has to prove that the bulk of its resources and power won't endlessly be locked into Iraq, despite the 2008 withdrawal agreement. The Iraqis are certainly ready for us to withdraw. At the same time, we have to reassure both Israel and our Arab allies that we will deter and counter any mounting threat from Iran, even as we engage Iran in broad talks to change the overall relationship. As Nixon and Kissinger did with China, the strategy should show how we are both reeling in the Iranians with prospects of improved relations, while deterring them.

You'll need extensive consultations before the speech. That doesn't mean your minions should go around asking everyone, "What should we do?" There's no quicker and surer way to lose confidence, especially among Arabs. If you don't come to them with the core of your strategy intact, they'll think you have no confidence in your own strategy. Give them your overall approach, then listen to them all—senators, Arabs, and Israelis. They'll tell you things you hadn't thought about enough. Include their suggestions, so long as they don't trash your basic strategy. Be confident, not cocky.

You'll want to set realistic expectations. Specific dates should be avoided, but you'll need to convey that you're looking at the duration of your first term for results to materialize. On a background basis, you can explain to the press and Congress that you expect the full transfer of war-making responsibility to Iraq to take two to three years, unrushed talks with Iran to take three to four years, and the Palestinian-Israeli political situation to gel over perhaps the same amount of time. (Of course, you'll have some secret successes to pull out of your back pocket as you go along.)

Then, you'll need to show that you're not simply waiting for Middle Eastern miracles. You'll pledge to go full speed ahead at home to develop alternative fuels and conservation programs to lower dependence on Gulf oil. You will unveil an action plan to make homeland security a reality rather than a political laugh line.

America can't be a serious country without personal safety and economic resiliency against terrorist attacks.

Your final punch line has to stress that the United States will not tumble into another diplomatic graveyard or military sinkhole in the Middle East on your watch. You won't fear to negotiate and make compromises in the mutual interest. You will put American power on the line for the best outcomes in the region. But either way, with or without peace in the region, you will prepare for the worst at home and ensure American security.

CHAPTER 13

NECESSITY, CHOICE, AND COMMON SENSE

Foreign policy is common sense, not rocket science. But it keeps getting overwhelmed by extravagant principles, nasty politics, and the arrogance of power. These three demons rob us of choice, which is the core of a commonsense foreign policy.

The United States is declining as a nation and a world power, with mostly sighs and shrugs to mark this seismic event. Astonishingly, some people do not appear to realize that the situation is all that serious. A few say it is serious and hopeless, and still others that it's serious but reversible. I count myself among those who think the situation is most serious yet within our capacity to reverse—if we're clear-eyed about the causes and courageous about implementing the cures.

As of now, in the view of many, the United States is becoming merely first among major powers, and heading to a power level between present-day China and our current still exalted position. This would be bad news for both the United States and the world. Were this to happen over time, it would leave nations without a leader to sustain world order and help solve international problems. No single country or group of countries, and no international institution could conceivably replace America in this role—and leaders the world over know this well.

The decline starts with weakening fundamentals in America. First among them is that we've allowed our economy, our infrastructure, our public schools, and our political system to deteriorate. The result is diminished economic strength, a less vital democracy, and a mediocrity of spirit. These conditions are not easy to reverse. A second reason for our decline is how ineffectively we have used our international power, thus allowing our and others' problems to grow and fester.

Our nation must attend to both problems, the former even more than the latter. But here I attend principally to fixing how we think about American power and make our foreign policy. The decision-making drill, though complex in substance, is a relatively straightforward process. Most foreign policy professionals understand it well: Find out what's really going on in other countries; figure out our problems and opportunities, the likely interplay of power, and what we can and can't accomplish to meet essential needs. Of course, professionals will argue among themselves and make mistakes. But both the arguments and the mistakes will be within reasonable bounds, and policy will be adjusted as events evolve. The problem is that we have often transformed this sensible procedure into farce and farce into tragedy.

Foreign policy is common sense, not rocket science. But it keeps getting overwhelmed by what I call the three demons: extravagant principles, nasty politics, and the arrogance of power. Time and again, these demons come to dominate or at least exercise disproportionate influence on governmental and public debate about foreign policy. They are persistent tormenting forces, not subject to the normal give-and-take. They rob officials of choice, which is the core of commonsense policy. The true believers and the cynics who employ these demons to their advantage are hard to counter with arguments based on reason or fact.

Once the demons grab hold of a policy, American leaders leap skyward, with wild descriptions of threats and assertions of bold and unattainable goals. At these moments, when chasms yawn between rhetoric and reality, our leaders commit their most tragic and costly mistakes—mistakes this country is no longer strong enough to afford.

The demons ensnare our leaders into thinking about what they "must" do, rather than about what they can do. They create seeming necessities or imperatives that rob us of choice and thus of the essence of our common sense. And common sense should be the basis of our foreign policy. It is, after all, what our leaders usually wind up resorting to after years of policy failures.

Common sense has to be our answer precisely because it is less about answers and more about questions, the very questions we have ordinarily run away from. It tells us not what to think about problems, but how to think about them systematically. Common sense is what rescued us at critical periods of our history. It is what saved our values from floating off and becoming empty dreams and focused them instead on concrete plans to reconstruct Western Europe and Japan after World War II. It's how we won the war in Asia after we lost the battle in Vietnam. It's how the Cold War was won without a war. It is hard to imagine any other approach that can possibly fit this new and bewildering world.

A return to common sense—pragmatic, problem-solving—won't be easy. Those possessed by the demons are much tougher fighters than the moderates who are constrained by the reasonableness of common sense. But common sense is worth the fight because it offers the best hope for using America's substantial power effectively, and because power is still the necessary means to solve problems in the international affairs of the twenty-first century.

THE FIRST STEP BACK to a realistic grasp of power is to face the fact that the United States is beginning to decline at home and abroad.

The bases of America's international power are U.S. economic competitiveness and political cohesion, and there should be little doubt at this point that both these bases are in decline. Many acknowledge and lament faltering parts here and there, but they avoid a frontal stare at the deteriorating whole. It is too depressing to do so, too much for most people to bear. The federal deficit is now projected at up to $1.85 trillion for fiscal year 2009, a three-fold increase over 2008, and, with the costs of medical care and social security skyrocketing, is likely to get even larger. The federal debt is already staggering as it tops $10 trillion. We are now the biggest debtor nation in history, and no nation with a massive debt has ever remained a great power. Our heavy industry has largely disappeared, having moved to foreign competitors, which has cut deeply into our ability to be independent in times of peril. Our public school students trail their peers in other industrialized countries in math and science. They can't compete in the global economy. Generations of Americans now, shockingly, read at the grade-school level, and know almost no history, not to mention no geography. They are simply not being educated to become guardians of a democracy.

These signals of decline have not inspired politicians to put the national good above partisan interests, or problem solving above scoring points. Republicans act like rabid attack dogs in and out of power, and treat facts like trash. Democrats seem to lack the decisiveness, clarity of vision, and toughness to govern. This tableau of domestic political stalemate begs for new leadership.

The nation that not so long ago outproduced the rest of the world in arms and consumer goods, the nation lionized and envied for its innovation, its can-do spirit, and its capacity to accomplish economic miracles, has become Katrina-ized and Iraq-ized—overwhelmed by the tasks it once performed with competence and relative ease. In many areas of public endeavor, we are now incompetent.

Conditions in many countries around the world have improved considerably, but in many other countries, they remain woeful and are getting worse. There is now a steady stream of internal conflicts and genocidal bloodlettings, a cascade of failed and failing states, whiffs of renewed nasty competition among great powers, a wildfire of international crime, worries about worldwide health pandemics, food shortages, environmental disasters, rampant religious extremism, and a ceaseless, deadly international terrorist threat.

The world is an almost unfathomable montage of primitivism amid unprecedented global plenty—of new riches in countries that have been poor for centuries, of great gulfs between rich and poor nations, and between the rich and the poor within them. It is as if almost half the world has reverted to a Hobbesian state of nature, where life is "nasty, brutish, and short," a pre-state condition where tribes devour each other in ritual rivalries. Meanwhile, most of the other half, including both mature and budding nations, now have to spend more of their own time and resources restoring their own shaky economies and can do little to help the drowning nations deal with their poverty or insecurity.

The real danger in this universe of primitivism and plenty is not of new wars or explosions among major states, or a world war, or even a nuclear war. It is the specter of nations drowning—constant drownings in the face of the steady uptick in terrorism from Somalia to India, civil wars, tribal and religious hatred, lawlessness, poverty amid extravagance, disease, environmental calamities, governmental incompetence, weak leadership and weak governments, cruelty, and indifference. Many nations are going under because they are simply unable to cope, and they will drag others down with them.

Managing these problems lies beyond the power of weak and poor states themselves. And they don't receive much succor from their neighbors and regional organizations. The United Nations helps with refugees, health, and the like, but its members don't seem eager to take on additional responsibilities. Nongovernmental organizations heroically make life more bearable for ordinary people in unbearable situations with irredeemable governments. Major powers like India, China, and Russia are not ready to lend others a hand, both because they are still evolving and because they lack the tradition of helping those less fortunate than themselves. Europe and Japan do help in various ways, consistent with sustaining their own high standards of living. Then again, only the starry-eyed expected more from these countries.

What is unexpected and tragic is that the international drownings are multiplying at the very moment of America's decline—when the one nation most likely and most able to help them can't do as much as it has done in the past. The tragedies take place when the United States can't prevent them with new Marshall Plans and new

NATOs. Nor have the foreign policies of our last two presidencies provided the world or the United States with much to applaud.

AFTER CONFRONTING THE REALITY of America's declining power, the next step toward redeeming our nation is to acknowledge and to confront the destructiveness of the demons. Principles, politics, and the arrogance of power have done—and will continue to do—great damage to our own interests.

Everyone wants to be on the side of advancing freedom and democracy and combating communism and terrorism. But that desire, in turn, opens the floodgates to setting unachievable objectives. No presidential appointee wants to be accused of making proposals that are perceived as weak or that will risk losing the next election for the president or his party. And no civilian feels comfortable in the wimpish position of telling the U.S. military that it can't do the job (especially when the military itself is loath to admit any shortcomings). Even when the demons don't win arguments outright, they triumph in fixing the direction of policy or in eliminating viable alternative policies.

Everyone who has served in government is familiar with these experiences. Everyone is also aware that how people behave in these demon-driven situations determines their reputations and how they will be portrayed—as liberals, wimps, non–team players, loose cannons—and that these characterizations usually last forever. Few have been punished in the government job market for being a conservative or a hawk. People don't like to talk about this, but it is the constant, private lament of Washington's professionals.

The demons, to be fair, are both a blessing and a curse. In a curious way, they derive even more power from this duality. Principles or ideals serve both to ennoble American causes and to produce excesses. There's a fine line between the good reasons for promoting democracy in Egypt, for example, and pushing leaders of that nation toward offering unrealistic and unwise political concessions to extremists. Politics is at once integral to the democratic process at home, as well as the cause of politicians acting against the national interest in order to win or stay in public office. While acting with confidence is good, it must fall short of the arrogant belief that you can do anything regardless of realities.

It takes courage to question whether a country is ready for democracy and whether Washington is pushing too hard. Doing so exposes the questioner to charges of callous indifference to American ideals and disrespect for foreign cultures. Such a person is readily tarnished as a nonbeliever in America or someone with prejudices against Arabs or Africans.

The problem with principles is not unique to ideals such as freedom and democracy. It applies to any set of principles that are elevated into dogma. These principles include anticommunism, antiterrorism, realism, and globalization as doctrines. They all lead to exaggerated goals and truncated means. They all divide the world into black and white, thus inhibiting the drawing of serious and subtle distinctions.

Whereas principles are waved around flagrantly, politics in foreign policy is more like the proverbial elephant in the room—the overwhelming presence that everyone pretends not to notice. It's there, and everyone is calibrating its effects. But mentioning it is taboo. No one wants to bring profane domestic politics onto the sacred turf of national security. Retired senior officials rarely even touch on the subject in their memoirs. Politics comes into the open only regarding nonsecurity matters like trade or immigration, which are generally recognized to be "political." Otherwise, it's a subject one talks about only with friends who are moving to Oklahoma.

A terrible price is paid for all this silence—the crucial political calculations being made about the political viability of a policy are never seriously examined. Participants in the debate generally make their own private assessments of the politics and build these into their policy recommendations. None of these judgments gets examined, and the consensus almost invariably points toward toughening policy, but often without persuasive reasons for doing so. Few are willing to appear not tough enough and risk not being invited to the next meeting. This silence also opens the door to exaggerating threats to justify the bolder counterthreats.

Though the following story is not exactly on point, it offers a telling glimpse at what politics does to ostensibly sane people in Washington. As I recall it, one day Senator Jacob Javits swooped me up from my desk into his office. "We've got to get right down to the floor," he barked. "There's an important issue I've got to deal with. The Postal Service has issued a stamp commemorating the first circus in America, and they've got it wrong. It wasn't in Clearwater, Florida, it was in Saratoga Springs, New York." When we arrived on the Senate floor, the scene was nothing short of miraculous. It was still morning, when senators appeared only to give pro forma speeches like noting the independence day of Estonia. For the only time in the two years I spent with this great senator from New York, there were upwards of fifteen senators present and debating with some vehemence where exactly the first circus had been performed in America: "No, it was in Clearwater, Michigan." "No, it was in Oshkosh, Wisconsin." And so on. Each pronounced in turn and then went on to name a city in his own state. Old Senator John Williams of Delaware heard the ruckus, wandered onto the floor, and sat at a desk with an amused smile. Finally, he asked for the floor, and intoned: "You are all wrong. The first American circus was not in Clearwater, Florida, or Saratoga Springs, or—It was in Philadelphia in 1776 when the Continental Congress declared itself to be the U.S. Congress." If Congress often resembles a circus, and it does, high-level meetings in the executive branch on U.S. foreign policy sometimes seem divorced from reality, flying like a kite with only a string of reason to link it to planet earth.

Along with politics and principles, the arrogance of power—the runaway confidence Americans have that they can do anything they put their minds to—also attends these meetings. It's our can-do spirit, our fierce national will, the American character, and it can vanquish even the laws of gravity.

Self-confidence in the face of challenges is a boon. But embracing unfavorable odds without the wherewithal is hubris. No official wants to say that the United States can't do something—or anything. A military officer risks his career if he points out that the forces approved by the secretary of defense are inadequate to the mission. He opens himself to the charge of lacking a can-do spirit. Ambassadors hesitate to argue against a demand by Washington for them to do some table-thumping abroad. Doing so would cast doubt on whether they have the stomach for the job. With exceptions, the premium is on thinking that we can do anything if we're tough and determined enough. Being weak is punished. Being tough and failing is not.

In a sense, Americans are driven to excess by the very qualities that make them so special: their self-confidence and their impulse to help others achieve a better and freer life. But these potent instincts allow the extremists in our midst to carry us away, to exaggerate the threats we face, to override sensible limits, and to narrow debate. The demons permit the extremists to ignore complexities and reduce arguments to potent simplicities.

These simplicities—good versus evil and toughness versus weakness—then play into the larger political arena, into the media's penchant for black-and-white drama, and into Congress's impulse to score points. For most of the last half century, the demons have prevailed in the internal battles over U.S. foreign policy.

AN EXAMINATION OF THE Cold War shows just how much damage the demons have caused. To be sure, there were impressive successes, such as President Harry Truman's phalanx of productive international institutions, President Richard Nixon's diplomatic triumphs around the world, and President George H. W. Bush's deft handling of the conclusion to the Cold War. But these have been overshadowed by many costly mistakes. Time and again, our leaders overrode their better judgment and gave in to the demons.

In the 1950s, Secretary of State John Foster Dulles persisted with his public theme of "rolling back" the Soviet empire in Eastern

Europe and stirred its captive peoples toward hopeless revolutions, even though he understood that President Eisenhower would never risk a world war for any of their causes. He was driven by his righteous anticommunism, which to him justified any cost.

President John F. Kennedy sent a few thousand Cuban expatriates to hell in the Bay of Pigs in 1961, though no one could explain to him beforehand how the mission could possibly succeed without the U.S. air cover he refused to provide. But he feared that if he called off the operations, Republicans would accuse him of wimping out. He was prepared to sacrifice the lives of the Cuban invaders for his own political reputation.

Presidents Kennedy and Lyndon B. Johnson warred in Vietnam based on the rationale that they were preventing the Sino-Soviet communist monolith from conquering first Asia and then the world. But from the late 1950s on, U.S. intelligence had firm evidence that no such monolith existed. In fact, the CIA knew there was a highly exploitable Sino-Soviet split. The two Democratic presidents and Nixon were all driven by anticommunism, fear of the political consequences of losing a war, and hubris about America's strengths.

After the Arab oil boycott in 1973 and its clear demonstration of America's dependence on a critical commodity from the world's most volatile region, Nixon and Henry Kissinger were sufficiently alarmed to create the position of "energy czar" and hold a conference on energy supplies. But, subsequently, neither they nor their successors did anything about that dependence, fearing the political costs of requiring Americans to pay extra taxes. The price of that inaction has been two wars, costly defense budgets, and enormous outflows of U.S. wealth.

President Jimmy Carter committed the blunder of thinking and stressing at the outset of his administration that the war on "isms" was over, despite increasing evidence that the Soviet Union was becoming more, rather than less, assertive. Carter was so convinced that our anticommunist ideology had been the major cause of crucial U.S. policy errors that he concocted a naive counter-ideology that weakened U.S. power and policy everywhere, as did his incredible passivity when Iran held fifty-two Americans hostage for 444 days. Carter was an unusual, if not unique, case of a president driven by a pacifist ideology.

President Ronald Reagan rightly helped the mujahideen oust Soviet forces from Afghanistan, but was so fixated on anticommunism that he completely missed the new terrorist threat there, just as he did in Lebanon after a terrorist attack killed 241 U.S. servicemen there in 1983. The terrorists surely got the message that he didn't want to tangle with them. Reagan was heading into difficult political winds and wanted to avoid popular backlash over further loss of American lives. And what but crazed anticommunism could have led to the Iran-contra affair: trading arms to Iran for American hostages in order to produce illegal monies to fund anticommunist contras in Nicaragua. His own adviser told him this was an "impeachable offense."

President George H. W. Bush, for all his sophistication in orchestrating the demise of the Soviet Empire and other ventures, committed serious errors of his own in the name of a realist foreign policy. First, he sided with Saddam Hussein against Iran by providing him with arms and intelligence, just as Reagan had. Then, Bush also looked the other way when Saddam began to threaten Kuwait. The U.S. ambassador to Iraq at that time, April Glaspie, had even gone so far as to tell Saddam: "We have no opinion on the Arab-Arab conflicts." Saddam surely took this as a green light to invade Kuwait. A hard line from Bush almost certainly would have stopped Saddam in his tracks.

Bush's mistakes all appear to have stemmed from an excessive devotion to an ideology of realism that undervalued values and overvalued Saddam's importance to the United States in containing Iran. Bush also essentially gave the green light to Serbian ethnic cleansing in Bosnia and Croatia when he went along with Secretary of State James Baker's line that Americans did not "have a dog in this fight." Stopping genocide did not square with Bush's realist conception of the national interest.

The list of President Bill Clinton's mistakes includes hesitation, backtracking, and inaction—the quintessential sins of liberal politicians. He seemed to operate on the erroneous assumption that he didn't need a foreign policy in the post–Cold War world, that a domestic and economic policy would suffice. During the presidential campaign, he promised to end the genocide in Bosnia, only to dally for three years before taking action. He tarried again before taking military action in Kosovo. But the worst, as Clinton admits, was his inaction in the face of the Rwandan genocide. Not only did he reject the Pentagon's modest plan to set up a safety zone on Rwanda's border, but he and his team also opposed sending additional UN troops. And he was plainly in search of domestic political redemption in his grossly overeager attempts to conclude major agreements with Yasir Arafat and Kim Jong Il in the last days of his administration. Throughout his years in office, Clinton didn't care much about what was going on overseas so long as he felt the American people didn't care either. He wasn't going to get out front and expose himself politically.

Which brings us to the fiascoes of President George W. Bush. He rushed blindly into the Iraq War without hard evidence and fought it for years without a clue—no information, no plans, just prideful boasting. He ignored an abundance of advice about these problems, all the while rejecting diplomacy. He frequently threatened Iran and North Korea, calling on them to halt their nuclear programs—or else—only to retreat at every turn. He ignored expert advice that his threats would fall on deaf ears. Now Pyongyang has nuclear weapons and Iran may not be far behind. Bush was captive to his own version of a new antiterrorist ideology, his utter devotion to the ultraright's mania for military force, and national arrogance, which he came to embody.

THE DEMONS SEEM PRIMED to haunt the country for the second decade of the twenty-first century. One can almost sense their giddy anticipation as they hear what American leaders, Democrats and Republicans alike, said about Afghanistan, Russia's military moves on Georgia, and other matters as 2009 came to an end.

Several unhappy truths can be predicted about Afghanistan: With little domestic opposition, the U.S. commitment will continue to grow and with the United States assuming increasing responsibility there, the Afghans will move little closer to self-reliance. So long as we are in Afghanistan and willing to make major commitments, we are not going to lose. But we are unlikely to get beyond stalemate, particularly because the Afghan government is itself corrupt and inefficient and ultimately the fight will be theirs. U.S. power can prevent defeat but not promise victory. That's up to the Afghans.

We need to deal with the threats and employ our power in Afghanistan and Pakistan in a more traditional and proven way. The question is whether we still have the leadership capacity to get ourselves together before we pour a great deal more lives and treasure into this war.

The Obama administration has said it would push for more effort on the part of the Afghans while providing more U.S. help and engaging in more diplomatic efforts. But what it has done is to further Americanize the war. While avoiding making an absolute commitment to "victory," as Obama called for in the presidential campaign, he now aims to ensure that Afghan territory won't be used by international terrorists. He still backs a massive buildup of Afghan forces to 134,000, a significant increase in U.S. economic assistance and American civilian presence, and troop increases that could take the U.S. total to 68,000, and perhaps beyond, by the end of the year.

The better foreign policy toward Afghanistan is what I call "surge, split, deter, contain the enemy, and withdraw." This alternative approach would set America's objective not at eliminating the international terrorist threat from Afghan soil but at significantly diminishing the threat and making it manageable by applying a different mix of U.S. power. Specifically, this policy would increase various forms of economic and military aid over the next couple of years; it would increase U.S.-backed counterinsurgency operations in order to give friendly Afghans confidence to fight for their future; and it would include a plan to withdraw U.S. combat troops after about three years. This approach would seek to split the enemy by granting the "moderates" some power in Afghanistan and by "renting" as many Taliban as possible. As Defense Secretary Robert Gates acknowledges, "There's some evidence that a fair number of the Taliban are not committed Islamists or extremists, and so they may be able to be wooed away." Washington would also need to come up with serious ways to punish the Taliban if it again supports international terrorism, and to contain extremist ambitions by building a coalition of neighbors—in other words, proven policies of deterrence and containment.

The most significant divergence between the Obama administration's foreign policy toward Afghanistan and this alternative actually starts with the goal itself. As long as the objective is to eliminate the threat of terrorism from Afghanistan, the American commitment will remain open-ended. The alternative policy is the only one to acknowledge the fact that no matter what happens in Afghanistan, America and its friends could just as well be attacked by terrorists operating right now from Pakistan, Yemen, and Somalia.

Another case in point is Washington's reaction in the summer of 2008 to Moscow's strategy first to bait and then to bloody Georgia over South Ossetia and Abkhazia, two Georgian provinces with close ties to Russia. After the Russians occupied the provinces and parts of Georgia as well, U.S. and European political leaders outdid themselves issuing vague warnings that Moscow should back off—though they never threatened military action or a boycott of Russian oil and gas. The threats were no more than wrist slaps: keeping Russia out of the World Trade Organization, dumping it from the G-8, and suspending its membership in various for-show NATO groups. Russian leaders eventually removed their troops from Georgia proper, but the rhetoric on both sides pointed toward overreach and trouble.

There are many additional opportunities for the demons. Our Sunni friends in Saudi Arabia could awaken one day to a revolution among their Shiites, and then see what happens to oil prices. President Obama will also have to muse on the prospect of civil war in a post-Castro Cuba and an Islamist revolution against the Mubaraks in Egypt. President Obama could also be the first to respond to a series of environmental disasters caused by global warming. And one can only imagine what the demons would demand in the event of new terrorist attacks on American soil.

THE DEMONS CREATE IDEOLOGICAL and political imperatives or necessities, and necessity admits of no serious discussion. It simply has to be. Necessity transforms centuries of Vietnamese culture and history into a manageable square on the strategic chessboard and makes victory in Iraq look as easy as the conquest of Panama. Necessity leads presidents and their advisers to establish dangerously unachievable goals that greatly exceed our power, that may or may not represent the wishes of the people they're intended to help, and that justify engulfing us in quagmires from which arrogance alone promises to extract us.

Defeating Hitler and Hirohito in World War II was a true necessity, but how best we could do this was up for debate. Containing Soviet communism was a true necessity, but the places and the means should have been debated and often weren't. Defeating the terrorists is a new true necessity, but how to distinguish among them and combat them needs to be freely examined, and that's already hard to do. Necessity removes choice.

The core problem here is not our democracy or our ideals or our power. It is ourselves. In part, leading Democrats and Republicans mishandle the politics of foreign policy. Most Democrats adhere to fundamental liberal beliefs about the value of negotiations and cooperation with other states. At the same time, however, they calculate that this will sound too soft to mainstream America. So, they seem to be torn between their beliefs and their politics, and they create the impression that they were for something before they were against it, and against it before they were for it. Democrats convey uncertainty about what they'll do, and the public senses this and then loses confidence in how they will manage national security.

By contrast, the Republicans exude nothing but conviction about being aggressive, standing up to any possible adversary, and painting the world in simple black-and-white. They are forever proclaiming that they will never allow America to be pushed around in the world. And though they have little regard for careful formulations of problems and difficulties, and though the public senses this as well, mainstream Americans appear to like the Republicans' conviction. Thus they have more confidence in the GOP than in the Democratic party on the handling of international affairs.

In part, moderates are reluctant to fight for the reasonable portrayal of problems and what we can do about them, or for choice, which they know to be the essence of a good foreign policy. The moderates know that good policy requires an open and honest review of the facts. They know that the effective use of power requires being able to push a range of buttons until some are found to work. Yet moderates don't fight for choice. Instead, they allow extremists to twist what is good and special about us—our ideals and our democratic politics—into a denial of choice. We cannot conduct an effective foreign policy if we allow necessity to crush choice.

THE FOREIGN POLICY COMMUNITY of experts and officials does not appear to be as alarmed as I am by the demons and the necessities they create. This community is less concerned with what derailed good policy in the past or what might do so again in the future, and more focused on what a good foreign policy should now be. That's fine—as long as that new policy both makes sensible use of American power and is capable of doing battle with the demons.

Most foreign policy experts are pushing for a new grand strategy to replace the old containment strategy. They are disposed toward big ideas and toward wedging all the pieces snugly together into that one big, neat theory. They're not enamored of loose ends or unintended consequences, which call their expertise into question. To their credit, most contribute value, perspective, and insights, although not without drawbacks.

The neoconservatives rightly remind Americans of the irredeemable and irreconcilable evil out there. But then they paint almost all foreign opponents (and some of their domestic ones as well) with a similar brush. They see past enemies such as Russia and China as future enemies as well. As for America's allies, particularly the European ones, the neoconservatives portray them as mostly worthless, lacking any useful military power and being averse to the use of force. Their list of enemies and unworthy allies is so extensive as to leave little room for allies and for the exercise of power other than the threat and use of military force. To the demons, they must look like best friends.

The reality is that neoconservatives will never be happy unless they are promoting some form of ideological warfare. Robert Kagan, for one, argues that instead of the old ideological clash between communism and democracy, there is a new one: democracy versus autocracy—the United States versus China and Russia.

But the leaders of Russia and China are not going around the world and proselytizing for their forms of government the way their communist predecessors did. Rather, Moscow is playing its old power games by trying to muscle its neighbors, but this time mostly with economic, rather than military, power. At this point China's leaders are interested almost solely in protecting themselves internally. The only preaching being done by these two autocracies is against American "unilateralism," and they do this to give themselves some elbow room for their own limited global concerns.

If there is anything approaching an ideological battle in the world today, it's between what other states perceive as American unilateralism and their own new sense of entitlement.

The realists, comfortable with power, rightly remind us to focus on our vital interests and not on all the world's problems. But they are often too impressed by power per se. Many of them were too eager to embrace Saddam, for all his sins and unpredictability, as a counterweight to Iran. Many now are eager to excuse the rough behavior of Russia and China as merely what big dictatorial nations do, and they have not paid much attention to how to use American power with failed or failing states or to address new transnational issues such as the environment. The realists continue to chafe at the value of values and the president's need to espouse them to sustain his foreign policy at home. Their realism is sometimes actually not realistic enough, and when it is not, they overlook both policy choices and areas that call for the application of power.

The liberal internationalists still exist today as an important element within the Democratic Party. Their most impressive contribution has been to keep reminding Washington of the need to cooperate with allies and negotiate with adversaries in almost all instances. And since the Vietnam War, they have been calling for new international institutions without being specific or practical about them, and they have been drifting toward softer and more unrealistic definitions of power. Formulating a strategy is difficult for them because it is mainly a call for more negotiations, more multilateral diplomacy, and less reliance on military power and force. To complicate matters further, when they come under great political pressure many of them appear to abandon these principles and become war hawks themselves, as happened with Iraq.

Interestingly, some of the liberal Democrats have joined with neoconservatives to form a new group that advocates a concert of democracies or some kind of institutional alliance to consolidate like-minded democracies. That sounds like a helpful project, and it might even be one if its advocates would demonstrate how they propose to corral the world's hundred or more democracies, including Botswana, Costa Rica, Peru, India, Israel, Mauritius, and South Africa, as well as most of Europe. Besides, they make little room in their concert for Russia and China, which aren't democracies but matter more than almost all those other democracies put together when it comes to diplomatic coalitions and power.

Then, of course, there are the globalizers who, to their credit, bear witness to the new centrality of economics, which the national security–oriented foreign policy clan traditionally ignores—out of ignorance. But globalizers still tend to overplay their hand by suggesting that economics will bring peace and democracy. Notoriously, they scant diplomatic and military choices.

George Kennan, who fathered the containment policy that most experts of today explicitly seek to emulate, would probably be first in line (along with the great Dean Acheson) to argue against a new unified field theory or a single holy grail for U.S. foreign policy. He would do so partly because he never believed in holy grails. But I think he would also point out that there are no defining contests among major powers of the kind that typically anchored and ordered overall strategies throughout history. I believe he would also say that no universal policy can satisfactorily encompass today's diversity of interests, the unprecedented range of problems, and the new power of the weak to resist the powerful.

A FOREIGN POLICY GROUNDED in common sense is probably the only approach that would satisfy Kennan's correct concerns. It is not a new holy grail. It does not revolve around great-power conflicts or the imagined absence of power conflicts. And it appreciates the diversity of the twenty-first-century world.

A commonsense policy, however, does not mean a seat-of-the-pants or ad hoc policy. It means an approach that allows our leaders to examine each situation on its own merits and link it to others when linking is justified by evidence and reason. Nor is such a policy rudderless; indeed, common sense insists that policy be grounded in strategy, priorities, and clear direction. Common sense also allows us to treat the strategy, the priorities, and the direction as guidelines, not straitjackets.

In contrast to all general policies, a commonsense policy is indeed untidy, but its positive attributes are indisputable. It comports with an untidy world and doesn't weed out facts that don't fit the theory. It offers most policy choices a fair hearing. It also accommodates our ideals, stressing achievability rather than posturing or threats. Common sense certainly doesn't preclude flexing our military muscles or using military force, either; deterrence, containment, and punishment remain crucial. But before the cannon fire begins, common sense demands a strict accounting of alternatives and of the probable consequences. A commonsense policy focuses on diplomatic and economic power but keeps military power ever in the background as the principal means of pressuring other nations and solving problems.

DEVISING A COMMONSENSE FOREIGN policy boils down to following five guidelines.

First, make America strong again by restoring our economic dynamism and pragmatic, can-do spirit.

Our resources and our will to help the world are the ultimate basis of our international power. If these deteriorate, our power abroad shrinks; it is as simple as that. To prevent this shrinkage will require giving far higher priority to energy independence, physical and human infrastructure, and homeland security.

Unless weaknesses in these areas are reversed, even our cleverest foreign policies will fail.

Second, understand clearly that mutual indispensability is the fundamental operating principle for power in the twenty-first century—meaning that the United States is the indispensable leader but needs equally indispensable partners to succeed. In other words, succeed together or fail apart.

Three points underpin this proposition: (1) the United States is the indispensable leader in the world and can't be replaced; (2) America's power to lead is not the power to dictate, but rather the power to solve major international problems; and (3) America needs to form coalitions with equally indispensable partners.

The aim here is not foolish multilateralism, but the creation of small and ad hoc power coalitions to solve particular problems.

Third, focus U.S. policy and the power coalitions that must be forged on addressing the greatest threats—terrorism, economic crises, nuclear proliferation, climate change, and global pandemics—and then just mind other threats as best you can. Policy and power can't work without clearly set priorities—priorities forever preached about but usually ignored.

The good news is that the United States and other key states have substantial common interests in these areas, which should facilitate cooperation. One problem will always be with us: the strain of forging common action when faced with unforeseen events and with events that seem to affect certain countries more than others.

Fourth, remember that international power works best against problems before, rather than after, they mature. This suggests the need to develop power coalitions early on to combat the agreed-upon and foreseeable threats.

It's fashionable to argue that power fares best in crises. That may be true in domestic affairs, but in international affairs, nations become so entrenched in their positions during hard times that in such moments war has historically been more likely than peace.

The issue, obviously, is to spur common action far in advance, putting the burden on the leading nation to provide the impetus and the proposed plans for dealing with international threats.

Fifth, realize that while the essence of power remains pressure and coercion based on a state's resources and international position, in other respects power isn't what it used to be.

The strong can't expect to command the weak as in the past; the weak now can resist, and do.

Traditional power doesn't work very well against today's problems—terrorism, poverty, tribal and religious conflicts, and climate change—because these exist mostly within nations rather than between nations. They're harder for power to reach or isolate.

Power today must focus more on riding economic and diplomatic tides rather than on initiating military storms. This, in turn, means that power will work more slowly now than before. Economics and diplomacy are slower processes than a military strike.

All this puts a premium on whether key nations will see the new contours of power correctly and will have the patience to give this power the time it needs to perform. Such patience is absolutely essential in the twenty-first century, because nations now are both too strong to simply roll over to demands by major powers and too weak to make concessions without time-consuming preparations.

There is a natural grouping in American politics to support such a commonsense approach—a union of traditional Republican Party realists such as Henry Kissinger and James Baker, and Truman-Acheson Democrats such as Joseph Biden and Sam Nunn. These two groups of realists have more in common with each other than with most members of their own political parties. While their party allies tend to focus on posturing for certain self-proclaimed ideals, these two groups of realists focus on solving problems with adversaries and allies alike by means of both diplomacy and military power. The union of these two groups of realists was exemplified in a recent book by Zbigniew Brzezinski, a Democrat, and Brent Scowcroft, a Republican. Basically, they demonstrated that there was a comfortable fusion of concern for power and interests, on the one hand, with long-term implementation of American values, on the other. These two groups are natural allies in thinking, though not in political action—and that's been the problem. They still stick largely to their own political kind.

THESE REALIST AND COMMONSENSE principles, which certainly can be adjusted by good minds, are not self-executing. They have to be fought for in the policy and political arenas, where the demons and their handlers have long presided. But moderate political leaders and moderate policy experts can pick up the cudgel of common sense, and win.

Moderates won the critical policy battles of the Cold War. You moderates, too, can wield common sense on the Senate floor, on television talk shows, and in White House meetings. You have lots going for common sense: its simplicity, its being every bit as much a part of the American character as the demons are of American politics, the record of historical tragedies caused by its neglect, and the poor state of affairs within America today.

Common sense has always been integral to American culture, and can be etched back into the political culture as well, if you keep pushing it and appealing to it. While some Americans have become addicted to extremism, most can be weaned away from this by emphasizing the practicality of common sense. You need to persist in asking questions, then asking them again: "What makes you think doing 'more' will work, and what if it doesn't?" "You've been taking this same position for years, and where has it gotten us?" "What do you mean by 'dire consequences'?" Make the advocates for the demons be specific. It will show how little they have behind their bluster.

Don't concede the values field to the extremists: You're all for freedom, democracy, and the American way, the practical way, the way we did it in Germany, Japan, and South Korea. Tell people you want a policy based on values—but one that works, and doesn't simply emit clouds of asphyxiating blue smoke. Make them answer questions about the viability of their approach. In debate, whoever has to answer the most questions loses. Be confident that Americans don't want to be led into blind alleys in which you will lose the national treasure. Stand your political ground.

Had our leaders rooted their arguments in common sense and fought for it, we would have been spared most of the policy horrors of the last fifty years. Most made no sense at the outset, or very soon thereafter. Where were the voices asking: "You mean to tell me that we should start a major war in Iraq without any idea of what we'll do when Saddam is gone, and before we finish the job in Afghanistan, home to Osama bin Laden?" And: "You mean to tell me that just because we made the mistake of committing our power to a losing and unjustified proposition, we should compound the mistake by sticking to it for another decade?" Where were the policy experts when it came to asking this: "If you say we can't accept Iran as a nuclear power, do we go to stop it and, if so, how do we square this with accepting nuclear-armed crazies like Pakistan and North Korea?" Or this: "Explain to me again how we pay the price for a better environment here without other countries doing the same, and still remain able to compete economically?" Or: "Can't we deal with adversaries the way we did with China, Russia or Libya, and make distinctions between adversaries and evil men?" Or: "Do we really think we'll be the first major power to pacify Afghanistan?" Or: "Do you actually believe that the terrorists will behave if we cure the world of poverty?" Or: "If what you're proposing doesn't work, what do we do next, and after that—and how is it all going to work?"

Yes, questions grounded in basic common sense get squelched for fear of political retribution and because our public debate has become overwhelmed by declarations of American principles and affirmations of American military might. But they get crushed mainly because moderates simply don't fight.

You'd think that by now all this would be obvious, and that we would have done something commonsensical about so many of what turned into our worst foreign policy misadventures.

You'd think by now presidents, more than anyone else, would have made that fight for common sense, if only to protect themselves from predictable failures.

You'd think that senators and congressmen would be ashamed and indeed mortified by their failure to seriously oversee fundamental decisions about war and our economy, our lives and treasure, or by their utter failure to have moved toward energy independence decades ago.

You'd think that our generals would speak out vigorously when their civilian superiors order them into battle without adequate resources and manpower, on missions that will only squander the brave lives under their charge.

You'd think that serious journalists would revolt when their networks and newspapers devote far more time and space to the weather, partisan shouting matches, lost mountain climbers, kidnapped children, O. J. Simpson, and the travails of Britney Spears than to major policy issues. You'd think their editors would know that our democracy can't function unless they help Americans sort out fact from fiction.

You'd think that community leaders across the nation, Republicans and Democrats alike, would be sickened by the absence of serious, intelligent public debate on domestic and national security issues. You'd think that these same people would be alarmed by the deterioration of public schools—and demand change.

You would think that our leaders would notice that our nation—the world's beacon of immigrants' dreams for centuries, and the only country that can maintain a semblance of world order—is heading toward winter.

Will it require an unmitigated disaster to wake us up? Or even then, will liberals and conservatives just continue to devour each other?

Every great nation or empire ultimately rots from within. We can already see the United States of America, our precious guarantor of liberty and security, beginning to decline in its leadership, institutions, and physical and human infrastructure, heading down the path to becoming just another great power, a nation barely worth fearing or following.

I was on precisely this depressing tear recently with a group of West Point cadets—and then I stopped in my rhetorical tracks. I said to them that this message of frustration and gloom was only half of what I really wanted to say to them. Yes, I wanted to send up flares signaling that we're losing our way and our power, and that we're in trouble. But even more important, I wanted to leave them with what is really driving me, with the strongest belief that we Americans are worth fighting for—both across the oceans and especially here at home. Don't doubt that we alone can provide the leadership to solve the international problems that will otherwise engulf us all. And for all America's faults, don't doubt that we remain the last best chance to create equal opportunity, hope, and freedom. But to restore all that is good and special about our beloved country and to rescue our power to solve problems will require something that has not happened in a long time: that pragmatists, realists, and moderates will unite and fight for their America.

ACKNOWLEDGMENTS

Of course, I cherish and thank those who have loved and nurtured me throughout my life and during the ordeal of this book. They gave me courage. But I thank equally those who have disagreed and argued with me. They made me smarter.

Alone at the top for both reasons is Judy Gelb, my life's love and chief critic, my all. She also conjured up the title of this book.

Then, surely, come my fantastic and beloved children—Adam, Caroline, and Alison—and their children: Sam, Max, Patrick, Jacob, and Will. They are a blessing.

Wiley, Tiggy, and Olivia always inspired and comforted me. I'm crazy about those kitties.

I am exceedingly lucky in my friends, many of whom double as brilliant diplomats and foreign policy thinkers: I write here of Fouad Ajami, Morton Abramowitz, Reginald Bartholomew, Richard Holbrooke, Winston Lord, Strobe Talbott, and Frank Wisner. They are all the best our nation has to offer. As friends do, they tried to save me from myself—alas, in vain. I must make special note of Lord, Bartholomew, Ajami, and Holbrooke for having waded through every draft with great care and wisdom. In particular, I owe a monumental intellectual and editorial debt to Win Lord.

My close friend Elliot (Skip) Stein also smartened the book throughout every draft.

Peter G. Peterson, soul mate and great American, did not contribute a single idea to this book, but provided instead the joys of friendship, good counsel, and constant and unjustified harassment. We had a phenomenal partnership during my ten years as president of the Council on Foreign Relations and my five subsequent years as a board fellow. Thanks, too, during all these Council years, to Maurice (Hank) Greenberg.

More appreciation for reading various drafts goes to another batch of close and brilliant friends: the legendary editor of *Foreign Affairs* James Hoge, Theodore Cross, Mary Cross, Michael Kramer, Robin McNeil, Peter Osnos, Richard Plepler, Richard Steadman, Julia Sweig, and Peter Tarnoff.

And for good attention to the chapter on economic power, I thank Samuel Berger, Benjamin Friedman, Jeffrey Garten, Timothy Geithner, Sebastian Mallaby, and C. Fred Bergsten.

And for good attention to the chapter on military power, I thank Bernard Trainor, William Nash, Robert Ferrell, Paul Greenwood, Jeffrey Harley, and Jeffrey Kendall.

My terrific colleagues at the Council on Foreign Relations almost always made life better and writing easier. Among them were Council President Richard Haass, Michael Peters, Jan Murray, Irina Faskianos, Anne Luzzatto, Nancy Bodurtha, Lisa Shields, Elise Carlson Lewis, Jan Mowder Hughes, Valerie Post, Leigh Gusts, David Kellogg, Patricia Dorff (also an excellent editor), Mark Hudson, Glen Goldman, Nidhi Sinha, Frank Alvarez, Phil Falcon, Charles Day, Deepak Trivedi, Robert Allende, and Roberto Osoria.

My deepest appreciation goes to the Mandelbaums, Michael and Anne. To Michael, for imparting his great knowledge and judgment. To Anne, for her sense of history and great skill with the English language.

Morton Janklow—literary agent extraordinaire, intellectual companion, and buddy—shepherded the enterprise and me from beginning to end.

No one could have a better editor than Tim Duggan. Smart, knowledgeable, assertive when necessary, and otherwise restrained.

He made the book better and better. Thanks as well to the entire team at HarperCollins, especially Allison Lorentzen, Susan Gamer, Tina Andreadis, and Kate Blum.

This book would not have been possible without the generous support of the Ford Foundation, the Anna-Maria and Stephen Kellen Foundation, the Peter G. Peterson Fund of the New York Community Trust, and the Starr Foundation. I saved for last my research associates at the Council on Foreign Relations: Tom Sullivan, Darren Geist, and Jeanne-Paloma Zelmati. They were the best in care and intelligence. I commend them to the world. Jeanne deserves special note for working with me through the hard final steps of the book. So does my other assistant at the Council, Kate Howell, for helping everything work right. So does my researcher from thirty years ago, Matthew Murray, who fortunately couldn't keep his hands off the final version of this book. And special thanks to Elva Murphy, my longtime secretary and chief organizer at the Council.

I see the faces and hear the voices of my mother and my father, immigrants from the Carpathian region of the Austro-Hungarian Empire, and I remember that day decades ago when they visited my overly large office in the Pentagon, saw the American flag behind my desk, and wept. I weep over their memory, and I weep for my country.

NOTES

LETTER TO OUR ELECTED PRINCE

ix   "gaze down from the summit . . . cruel fate": Niccolò Machiavelli, *The Prince and the Discourses* (New York: Random House, Inc., 1950), 4.

INTRODUCTION

xiii   "[I]t was reserved for Augustus" to "the most formidable Barbarians": Edward Gibbon, *The Decline and Fall of the Roman Empire*, Vol. 1 (New York: Modern Library, 2005), 1–2.

CHAPTER 1: THE REVOLUTION IN WORLD POWER

7   "Diplomacy without arms is like a concert without a score": Garrett Mattingly, *Renaissance Diplomacy* (London: Jonathan Cape, 1955), 155.

7   "Shiftiness and inconstancy . . . each state against all": Ibid., 156.

7   "Nor in the decade . . . business of kings": Ibid., 134.

14   "a pitiful, helpless giant": Richard Nixon, "Address to the Nation on the Situation in Southeast Asia," Richard Nixon Library and Birthplace Foundation (April 30, 1970).

18   "axis of evil"—Iraq, Iran, and North Korea: George W. Bush, "The President's State of the Union Address: 2002," Office of the White House Press Secretary (January 29, 2002).

23   Stanley Hoffmann . . . international law, and norms: Stanley Hoffmann, "International Organization and the International System," *International Organization*, 24:3 (Summer 1970).

23   Joseph Nye and Robert Keohane . . . added to these restraints: Robert Keohane and Joseph Nye, *Power and Interdependence: World Politics in Transition* (Boston: Little Brown, 1977).

CHAPTER 2: WHAT POWER IS, AND WHAT POWER ISN'T

26   "I can call spirits" . . . "when you do call for them?": William Shakespeare, *Henry IV: Oxford School Shakespeare* (Oxford: Oxford University Press, 2002), 63.

30   *potere*, "to be able to": J. A. Simpson and E. S. C. Weiner (eds.), *The Oxford English Dictionary*, 2nd ed., Vol. 12 (Oxford: Clarendon, 1989), 259.

30   According to . . . Middle English in 1297 as *poer*: Ibid.

31   "A prince . . . concern to one who commands": Machiavelli, 53.

32   "great beef-fed men, red-faced and red-coated": George Orwell, *Burmese Days* (New York: Harcourt, Brace, and World, 1950), 5.

32   "*A* has power over *B* . . . *B* would not otherwise do": Robert Dahl, "The Concept of Power," *Behavioral Science*, 2:3 (July 1957), 202–3.

CHAPTER 3: POWER IN THE AMERICAN CENTURY

47–48 Kennan gave the basic American policy of containment . . . the entire byline: X, "The Sources of Soviet Conduct," *Foreign Affairs* (July 1947).

48   "impervious to logic of reason": Telegram, George Kennan to George Marshall ["Long Telegram"], February 22, 1946. Harry S. Truman Administration File, Elsey Papers, http://www.trumanlibrary.org.

48   "There is no reason . . . superior strength at every turn": Walter Isaacson and Evan Thomas, *The Wise Men: Six Friends and the World They Made* (New York: Simon and Schuster, 1997), 374.

48   "the manipulation of our political . . . military strength": Ibid., 375.

49   "The dog . . . mutual relationship": Ibid., 447.

49   "We must refrain . . . success in terms of years": Ibid., 470.

49   "You cannot sit down with [the Soviets]": Robert L. Beisner, *Dean Acheson: A Life in the Cold War* (New York: Oxford University Press, 2006), 62.

49   "situations of strength": Ibid., 149–50.

49   "You can dam [communist ideology] up . . . you can't argue with it": Ibid., 153.

50   "recognize facts": Ibid., 156.

52   "moving again": "John Kennedy," The White House, http://www.whitehouse .gov/history/presidents/jk35.html.

57   "major instrument": Leslie H. Gelb and Paul C. Warnke, "Security or Confrontation: The Case for a Defense Policy," *Foreign Policy* (1970).

59   "an evil empire": Ronald Reagan, "Speech to the House of Commons" (June 8, 1982) http://www.fordham.edu/halsall/mod/1982reagan1.html.

61   vital to U.S. interests: Terry L. Deibel, "Why Reagan Is Strong," *Foreign Policy*, 62 (Spring 1986).

62   "a new world order": George H. W. Bush, "Address to a Joint Session of Congress and the Nation" (September 11, 1990) http://www.sweetliberty .org/issues/war/bushsr.htm.

67   Bush had warned Iraq . . . State of the Union Address: George W. Bush, "The President's State of the Union Address: 2002."

68   "One is to threaten them . . . carrots and sticks": Joseph S. Nye, "Soft Power: The Means to Success in World Politics," speech at the Carnegie Council (April 13, 2004) http://www.cceia.org/resources/ transcripts/4466.html.

68–69 Soft power is . . . "achievement of those values": Joseph S. Nye, "Soft Power and Leadership," *Compass: A Journal of Leadership* (Spring 2004).

69   "co-opts people rather than coerces them": Joseph S. Nye , *Soft Power: The Means to Success in World Politics*, (New York: PublicAffairs, 2004), 5.

69   "Soft power rests on the ability to shape the preferences of others": Ibid.

69   "Credibility and legitimacy are what soft power is all about": Joseph S. Nye, "The Velvet Hegemon," *Foreign Policy* (May/June 2003).

69   "Military prowess and competence . . . 'than a weak horse'": Joseph S. Nye, "Think Again: Soft Power," *Foreign Policy* (March 1, 2006).

69–70 "Legitimacy is central . . . carrots and sticks": Richard Armitage and Joseph S. Nye, "CSIS Commission on Smart Power" (Washington, D.C.: Center for Strategic and International Studies, November 6, 2007), 6.

70   "the dominant power . . . demonstrations of will": Charles Krauthammer, "In American Foreign Policy, a New Motto: Don't Ask. Tell," *Time* (February 26, 2001).

70   "hopelessly utopian": Charles Krauthammer, "The Unipolar Moment Revisited—United States World Dominance," *The National Interest* (Winter 2002).

70   "What stability we do enjoy . . . deterrent threat of the United States": Charles Krauthammer, "Democratic Realism: An American Foreign Policy for a Unipolar World," Speech at the American Enterprise Institute (February 10, 2004) http://www.aei.org/publications/pubID.19912,filter.all/pub_detail.asp.

70   "Most Americans . . . impressed the world": Charles Krauthammer, "The Hundred Days," *Time* (December 12, 2001).

70   "com[e] not from love . . . newfound respect": Ibid.

70   "We will let them hold our coats, but not tie our hands": Charles Krauthammer, "The Axis of Petulance," *The Washington Post* (March 1, 2002).

71   "You will be not only disarmed but dethroned": Charles Krauthammer, "The Obsolescence of Deterrence," *The Weekly Standard*, 8:13 (December 9, 2002).

71   "classical deterrence": Krauthammer, "Democratic Realism."

71   "The new unilateralism . . . balancer of last resort": Krauthammer, "The Unipolar Moment Revisited."

71   "will cooperate with us . . . distant third": Ibid.

71   Thus the twenty-first century bears witness . . . democratic values and interests: Ivo Daalder and Robert Kagan, "The Next Intervention," *The Washington Post* (August 6, 2007), A17.

CHAPTER 4: THE NEW PYRAMID OF WORLD POWER

74   "Even the experts. . . out there": Personal conversation.

76   An intelligence study in late 2008: National Intelligence Council, "Global Trends 2025: A Transformed World" (November 2008), http://www.dni.gov/nic/PDF_2025/2025_Global_Trends_Final_Report.pdf.

CHAPTER 5: STRATEGY AND POWER: MUTUAL
INDISPENSABILITY

94   "We will bury you": "We Will Bury You!" *Time* (November 26, 1956).

96   "[H]istory demonstrates . . . rapidly than the opponent": Henry Kissinger, *Nuclear Weapons and Foreign Policy* (New York: Harper, 1957), 21–22.

104   "pitiful, helpless giant": Nixon, "Address to the Nation on the Situation in Southeast Asia."

105   "conspicuous successes": Henry Kissinger, "Kissinger's Visits to Beijing and the Establishment of the Liaison Offices, January 1973–May 1973," Memorandum of Conversation (January 3, 1973), http://www.state.gov/documents/organization/100320.pdf.

113   "As the history of this century . . . next century": "Excerpts from the Announcement on the Cabinet," *The New York Times* (December 6, 1996).

113   "If we have to use force . . . further into the future": Bob Herbert, "In America; War Games," *The New York Times* (February 22, 1998).

114   "our shared task, with the help of friends from around the world": "Excerpts from the Announcement on the Cabinet."

114   "no other power or bloc . . . can be achieved": Brent Scowcroft, "The Dispensable Nation?" *International Herald Tribune* (August 7, 2007) http://www.nationalinterest.org/Article.aspx?id=14778.

115   "Success will be less . . . the people in between": "Remarks as Delivered by Secretary of Defense Robert M. Gates," Washington, D.C. (October 10, 2007) http://www.defenselink.mil/speeches/speech.aspx?speechid=1199.

116   "we have to be willing to be persuaded by others": "John McCain's Foreign Policy Speech," *The New York Times* (March 26, 2008).

118   "only a united front . . . nuclear aspirations": "Defense Secretary Gates Says International Effort Needed to Curb Iran's Nuclear Ambitions," Associated Press, (October 16, 2007).

CHAPTER 6: INTELLIGENCE AND POWER

123   "[F]or knowing afar off . . . remedy to be found": Machiavelli, 11.

125   "I looked the man . . . a sense of his soul": George W. Bush, "Press Conference by President Bush and Russian Federation President Putin" (June 16, 2001) http://www.whitehouse.gov/news/releases/2001/06/20010618.html.

125   "You two did visit . . . didn't you?": L. Fletcher Prouty and Oliver Stone, *JFK: The CIA, Vietnam, and the Plot to Assassinate John F. Kennedy* (New York: Citadel, 1996), 262.

129   the 2007 National Intelligence Estimate: "National Intelligence Estimate, Prospects for Iraq's Stability: A Challenging Road Ahead," Office of the Director of National Intelligence (January 2007).

130   "The greatest threats . . . distribution of power": Condoleezza Rice, "Transformational Diplomacy," speech at Georgetown University (January 18, 2006).

CHAPTER 7: U.S. DOMESTIC POLITICS AND POWER

138   "timeframe for a withdrawal . . . as soon as possible": "Interview with Iraqi Leader Nouri Al-Maliki," *Spiegel* (July 19, 2003).

138 "time horizon": Dan Eggen and Michael Abramowitz, "U.S., Iraq Agree to Time 'Horizon,'" *Washington Post* (July 19, 2008), A1.

147 Walter Cronkite . . . could not be won: Milton J. Bates, Lawrence Lichty, Paul Miles, Ronald H. Spector, and Marilyn Young (eds.), *Reporting Vietnam, Part One: American Journalism 1959–1969* (New York: Library of America, 1998), pp. 581–82.

CHAPTER 8: MILITARY POWER

161 "A prince should have . . . to one who commands": Machiavelli, 53.

161–62 Yet, in the less than two . . . of the Cold War: Richard F. Grimmett, "Instances of Use of United States Armed Forces Abroad 1798–2007," Congressional Research Service Report for Congress (January 14, 2008).

164 "collaborative and cooperative relationships": Josh White, "Gates Sees Terrorism Remaining Enemy No. 1," *The Washington Post* (July 31, 2008), A1.

165–66 The price tag . . . Afghan and Iraq wars: "Fiscal Year 2010 Budget Request: Summary Justification," United States Department of Defense, May 2009.

177–78 Ronald Reagan . . . called their deployment "vital": Deibel.

CHAPTER 9: ECONOMIC POWER

188 Private investors . . . value of the ruble: Andrew E. Kramer, "Russia Stock Market Fall Is Said to Imperil Oil Boom," *The New York Times* (September 13, 2008), A10.

191 As of December 2007 . . . foreign exchange reserves: Brad Setser, "China: Creditor to the Rich," *China Security*, Vol. 4, No. 4, Autumn 2008, pp. 17–23, World Security Institute.

192 sages like Thucydides . . . especially to sustain armies: T. E. Wick (ed.), *The Peloponnesian War, Thucydides* (New York: Random House, 1982).

195 "to develop programs . . . national security interest": "Department 'Gets Down to Business' by Assisting U.S. Firms, Promoting American Exports," *Department of State Dispatch*, 29 April 1991, 306–9, http://www.disam.dsca.mil/pubs/INDEXES/Vol%2013_4/Dept%20Gets%20Down%20to%20Business%20by%20Asst%20U%20S%20Firms%20Promoting%20American%20Exports.pdf.

196 In most of the world . . . Machiavelli and Clausewitz: Adam Smith, *The Wealth of Nations* (New York: Bantam Classics, 2003); David Ricardo, *Principles of Political Economy and Taxation* (New York: Cosimo Classics, 2006).

196 Russian leaders are now . . . in Europe and elsewhere: "IMF Data Mapper: World Economic Outlook, April 2009," International Monetary Fund, http://www.imf.org/external/datamapper/index.php; "Country Statistical Profiles 2009," Organisation for Economic Co-operation and Development, http://stats.oecd.org/Index.aspx?DataSetCode=CSP2009.

199 China's gross domestic product . . . if not much longer: "Clipping the Dragon's Wings," *The Economist* (December 19, 2007).

199–200 China's biggest customer . . . $1 trillion since 1995): "Trade Profiles," World Trade Organization Statistical Database, http://stat.wto.org/

CountryProfile/WSDBCountryPFHome.aspx?Language=E; "Direction of Trade Statistics." International Monetary Fund, http://www.imf statistics.org/DOT.

201   Japan broke major taboos . . . the First Gulf War: Philip Shenon, "Actions of Japan Peacekeepers in Cambodia Raise Questions and Criticism," *The New York Times* (October 24, 1993); Andrew Bennett, Joseph Lepgold, and Danny Unger, "Burden-Sharing in the Persian Gulf War," *International Organization*, 48:1 (Winter 1994), pp. 39–75.

203   Although economists argue . . . $700 billion in 2007: "U.S. International Transactions, 1960–present," Bureau of Economic Analysis, http://www .bea.gov.

208   In fiscal year 2006 . . . about 150 countries: "Foreign Military Financing Account Summaries," Department of State, http://www.state.gov/t/pm/ ppa/sat/c14560.htm. "International Military Education and Training Account Summaries," Department of State, http://www.state.gov/t/pm/ ppa/sat/c14562.htm.

210   In 2006, U.S. ODA . . . Japan third at $11.2 billion: "Development Aid at its Highest Level Ever in 2008," Organisation for Economic Co-operation and Development (March 30, 2009); "Query Wizard for International Development Statistics," Organisation for Economic Co-operation and Development, http://stats.oecd.org/qwids/.

211   Washington has imposed sanctions . . . under U.S. pressure: Gary Hufbauer, "Sanctions-Happy USA," Peterson Institute, Policy Brief 98-4 (July 1998), http://www.iie.com/publications/pb/pb.cfm?ResearchID=83.

215   Official development assistance . . . the last five years: "Query Wizard for International Development Statistics," Organisation for Economic Co-operation and Development, http://stats.oecd.org/qwids/.

215   By contrast . . . $160 billion annually: "U.S. Net International Investment Position at Yearend 2008," Bureau of Economic Analysis, http:// www.bea.gov/newsreleases/international/intinv/intinvnewsrelease.htm.

216   "Not only the wealth . . . prosperity of manufactures": Henry Cabot Lodge (ed.), *The Works of Alexander Hamilton* (Federal Edition), 12 vols. (New York: Putnam, 1904).

CHAPTER 10: STAGE-SETTING POWER

222   not one contained a clear . . . threatening military action: "Saddam Hussein's Defiance of UNSCRs," U.S. Department of State, March 20, 2003, http://www.state.gov/p/io/rls/fs/2003/18850.htm.

222   Indeed Bush . . . the day of the attack approached: "Interview on NBC's Meet the Press with Tim Russert," U.S. Department of State (October 20, 2002), http://www.state.gov/secretary/former/powell/ remarks/2002/14495.htm.

227   "deals with the influence . . . inter-cultural communications": "What Is Public Diplomacy?" Fletcher School, Tufts University, http://fletcher .tufts.edu/murrow/public-diplomacy.html.

227   "American traditions . . . as simple as that": Philipp S. Muller, *Unearthing the Politics of Globalization* (Berlin: Lit Verlag, 2004), 64.

CHAPTER 11: FOREIGN POLICY POWER

238 Their report anticipated . . . dealing with them: Rachel Bronson, Edward P. Djerejian, Andrew Scott Weiss, and Frank G. Wisner, "Guiding Principles for U.S. Post-Conflict Policy in Iraq (Report of an Independent Working Group)," Council on Foreign Relations (January 1, 2003).

242 at least for . . . economic parity with us: "Clipping the Dragon's Wings."

247 Specifically, they told Moscow . . . and Central America: Henry Kissinger, *Diplomacy* (New York: Simon and Schuster, 1994), 716–17.

251 "bear full responsibility": "China Demands U.S. Apology, Allows Access to Crew," *Asian Political News* (April 9, 2001).

251 "nothing to apologize for": "Powell: 'We Regret' Chinese Pilot's Loss," *CNN Online* (April 4, 2001).

251 Then, Bush wrote . . . forced to land: "Diary of the Dispute," *BBC News* (May 24, 2001), http://news.bbc.co.uk/2/low/asia-pacific/1270365.stm.

251 "blatant aggression": Alan Cowell, "Iran Cites 'Aggression' in Case of Captured British Personnel," *The New York Times* (March 25, 2007).

251 "different phase": Alan Cowell, "Blair Pushes Iran for Release of Captives," *The New York Times* (March 28, 2007).

251 "will not happen again": Peter Walker and Mark Oliver, "Iran to Release British Sailors Tomorrow," *The Guardian* (April 4, 2007).

CHAPTER 12: U.S. POLICY AND POWER IN THE MIDDLE EAST

267 And most interestingly . . . peaceful nuclear technology: Nicholas D. Kristof, "Iran's Proposal for a 'Grand Bargain,'" *The New York Times* online, "On the Ground" blog (April 28, 2007), http://kristof.blogs.nytimes.com/2007/04/28/irans-proposal-for-a-grand-bargain/.

267 "were trying to put too much on the table": Glenn Kessler, "2003 Memo Says Iranian Leaders Backed Talks," *The Washington Post* (February 14, 2007), A14.

273 "hold onto their dreams": Richard N. Haass and George J. Mitchell, "Learning from Success," *International Herald Tribune* (May 7, 2007).

CHAPTER 13: NECESSITY, CHOICE, AND COMMON SENSE

280 The federal deficit . . . 2009: Jeanne Sahadi, "Deficit Estimate: Up to $1.85 trillion," CNNMoney Online (March 20, 2009).

282 "nasty, brutish and short": Thomas Hobbes, *Leviathan: Or, The Matter, Forme & Power of a Commonwealth, Ecclesiasticall and Civill* (Lanham, MD: University Press, 1904), 84.

288 "impeachable offense": "The Iran-Contra Affair 20 Years On: Documents Spotlight Role of Reagan, Top Aides," National Security Archive, Electronic Briefing Book No. 210, http://www.gwu.edu/~nsarchiv/NSAEBB/NSAEBB210/index.htm.

288 "We have no opinion . . . Arab-Arab conflicts": Gary Donaldson, *America at War Since 1945: Politics and Diplomacy in Korea, Vietnam, and the Gulf War* (Westport, CT: Greenwood, 1996), 149.

288 "have a dog in this fight": R. W. Apple Jr., "Clinton's Positive Thinking: The NATO Alliance Is Alive and Kicking," *The New York Times* (April 26, 1999). Thom Shanker, "Gates Gives Rationale for Expanded Deter-

rence," *The Washington Post* (October 29, 2008), A12.

294 Robert Kagan . . . China and Russia: Robert Kagan, "Why Should Democracy Be Shy?" *Newsweek* (June 9, 2008).

295 Some of the liberal . . . like-minded democracies: G. John Ikenberry and Anne-Marie Slaughter, "Forging a World of Liberty under Law: Final Report of the Princeton Project on National Security," Princeton Project on National Security (September 27, 2006).

299 a recent book by . . . a Republican: Zbigniew Brzezinski and Brent Scowcroft, *America and the World: Conversations on the Future of American Foreign Policy* (New York: Basic Books, 2008).

SELECTED BIBLIOGRAPHY

BOOKS

Acheson, Dean. *Present at the Creation: My Years in the State Department* (New York: Norton, 1987).

Ajami, Fouad. *The Arab Predicament: Arab Political Thought and Practice since 1967* (Cambridge: Cambridge University Press, 1992).

———. *Dream Palace of the Arabs: A Generation's Odyssey* (New York: Pantheon, 1998).

Albright, Madeleine. *Memo to the President Elect: How We Can Restore America's Reputation and Leadership* (New York: HarperCollins, 2008).

Bacevich, Andrew J. *The Limits of Power: The End of American Exceptionalism* (New York: Metropolitan, 2008).

Baker, James, III. *Work Hard, Study . . . and Keep Out of Politics! Adventures and Lessons from an Unexpected Public Life* (New York: Putnam Adult, 2006).

Barber, Benjamin R. *Fear's Empire: War, Terrorism, and Democracy* (New York: Norton, 2003).

Beisner, Robert L. *Dean Acheson: A Life in the Cold War* (New York: Oxford University Press, 2006).

Bergsten, C. Fred, and Lawrence B. Krause (eds.). *World Politics and International Economics* (Washington, D.C.: Brookings Institution, 1975).

Blechman, Barry M., and Stephen S. Kaplan. *Force without War: U.S. Armed Forces as a Political Instrument* (Washington, D.C.: Brookings Institution Press, 1978).

Bobbitt, Philip. *Terror and Consent : The Wars for the Twenty-First Century* (New York: Knopf, 2008).

Brzezinski, Zbigniew. *Power and Principle: Memoirs of the National Security Advisor, 1977–1981* (New York: Farrar, Straus, and Giroux, 1983).

Brzezinski, Zbigniew, and Brent Scowcroft. *America and the World: Conversations on the Future of American Foreign Policy* (New York: Basic Books, 2008).

Clifford, Clark M., with Richard Holbrooke. *Counsel to the President: A Memoir* (New York: Random House, 1991).

Chollet, Derek, and James Goldgeier. *America between the Wars: From 11/9 to 9/11* (New York: PublicAffairs, 2008).

De Grazia, Sebastian. *Machiavelli in Hell* (New York: Vintage, 1989).

Destler, I. M., Leslie H. Gelb, and Anthony Lake. *Our Own Worst Enemy: The Unmaking of American Foreign Policy* (New York: Simon and Schuster, 1984).

Dilenschneider, Robert L. *Power and Influence: The Rules Have Changed* (New York: McGraw-Hill, 2007).

Ferguson, Niall. *Empire: The Rise and Demise of the British World Order and the Lessons for Global Power* (New York: Basic Books, 2004).

Flanagan, Stephen J., and James A. Schear (eds.). *Strategic Challenges: America's Global Security Agenda* (Washington, D.C.: National Defense University Press, 2008).

Friedman, Thomas L. *Hot, Flat, and Crowded: Why We Need a Green Revolution—and How It Can Renew America* (New York: Farrar, Straus and Giroux, 2008).

Fromkin, David. *A Peace to End All Peace: The Fall of the Ottoman Empire and the Creation of the Modern Middle East* (New York: Holt, 1989).

Fukuyama, Francis. *America at the Crossroads: Democracy, Power, and the Neoconservative Legacy* (New Haven, CT: Yale University Press, 2006).

———. *The End of History and the Last Man* (New York: Avon, 1992).

Gibbon, Edward. *The Decline and Fall of the Roman Empire*, Vol.1 (New York: Modern Library, 2003).

Haig, Alexander M., Jr. *Caveat: Realism, Reagan, and Foreign Policy* (New York: Macmillan, 1984).

Harper, John Lamberton. *American Machiavelli: Alexander Hamilton and the Origins of U.S. Foreign Policy* (Cambridge: Cambridge University Press, 2004).

Holbrooke, Richard. *To End a War* (New York: Random House, 1998).

Hoffmann, Stanley. *Gulliver's Troubles; Or, the Setting of American Foreign Policy* (New York: McGraw-Hill, 1968).

———. *Primacy or World Order: American Foreign Policy since the Cold War* (New York: McGraw-Hill, 1978).

Huntington, Samuel P. *The Clash of Civilizations and the Remaking of World Order* (New York: Simon and Schuster, 1996).

———. *Who Are We: The Challenges to America's National Identity* (New York: Simon and Schuster, 2004).

Irons, Peter H. *War Powers: How the Imperial Presidency Hijacked the Constitution* (New York: Holt, 2005).

Kagan, Robert. *Dangerous Nation: America's Place in the World from Its Earliest Days to the Dawn of the 20th Century* (New York: Knopf, 2006).

———. *Of Paradise and Power: America and Europe in the New World Order* (New York: Knopf, 2003).

———. *The Return of History and the End of Dreams* (New York: Knopf, 2008).

Kechichian, Joseph A., and R. Hrair Dekmejian. *The Just Prince: A Manual of Leadership* (London: Saqi, 2003).

Keegan, John. *A History of Warfare* (New York: Vintage, 1993).

Kennedy, Paul. *Grand Strategies in War and Peace* (New Haven, CT: Yale University Press, 1991).

———. *The Rise and Fall of the Great Powers* (New York: Vintage, 1987).

Keohane, Robert, and Joseph Nye. *Power and Interdependence: World Politics in Transition* (Boston: Little Brown, 1977).

Khanna, Parag. *The Second World: Empires and Influence in the New Global Order* (New York: Random House, 2008).

Kindleberger, Charles P. *World Economic Primacy: 1500–1990* (Oxford: Oxford University Press, 1996).

Kissinger, Henry. *Diplomacy* (New York: Simon and Schuster, 1994).

———. *Necessity for Choice: Prospects of American Foreign Policy* (Westport, CT: Greenwood, 1984).

————. *Nuclear Weapons and Foreign Policy* (New York: Harper, 1957).

————. *A World Restored* (New York: Grosset and Dunlap, 1964).

Leffler, Melvyn P. *For the Soul of Mankind: The United States, the Soviet Union, and the Cold War* (New York: Hill and Wang, 2007).

Leffler, Melvyn P., and Jeffrey W. Legro. *To Lead the World: American Strategy after the Bush Doctrine* (Oxford: Oxford University Press, 2008).

Lord, Carnes. *The Modern Prince: What Leaders Need to Know Now* (New Haven, CT: Yale University Press, 2003).

Machiavelli, Niccolò. *The Prince and the Discourses* (New York: Random House, 1950).

Mandelbaum, Michael. *The Case for Goliath: How America Acts as the World's Government in the 21st Century* (New York: PublicAffairs, 2005).

Masters, Roger D. *Fortune Is a River: Leonardo da Vinci and Niccolò Machiavelli's Magnificent Dream to Change the Course of Florentine History* (New York: Plume, 1998).

Mattingly, Garrett. *Renaissance Diplomacy* (Baltimore, MD: Penguin, 1955).

Meltzer, Allan H. *Keynes's Monetary Theory: A Different Interpretation* (Cambridge: Cambridge University Press, 1988).

————. *Money, Credit, and Policy (Economists of the Twentieth Century)* (Northampton, MA: Edward Elgar, 1995).

Morgenthau, Hans J. and Kenneth W. Thompson. *Politics Among Nations: The Struggle for Power and Peace*, 5th ed., rev. (New York: Alfred A. Knopf, 1978)

Neustadt, Richard E. *Presidential Power: The Politics of Leadership, with Reflections on Johnson and Nixon* (New York: Wiley, 1976).

Nicolson, Harold. *Diplomacy*, 3rd ed. (New York: Oxford University Press, 1969).

Nye, Joseph S. Jr., *The Paradox of American Power: Why the World's Only Superpower Can't Go It Alone* (Oxford: Oxford University Press, 2002).

————. *The Powers to Lead* (Oxford: Oxford University Press, 2008).

————. *Soft Power: The Means to Success in World Politics* (New York: PublicAffairs, 2004).

————. *Understanding International Conflicts*, 6th ed. (Essex, UK: Longman, 2006).

Schlesinger, Arthur M., Jr. *War and the American Presidency* (New York: Norton, 2004).

Shultz, George. *Turmoil and Triumph: Diplomacy, Power, and the Victory of the American Ideal* (New York: Scribner, 1995).

Sloan, Stanley R., Robert G. Sutter, and Casimir A. Yost. *The Use of U.S. Power: Implications for U.S. Interests* (Washington, D.C.: Institute for the Study of Diplomacy, 2004).

Sorensen, Theodore C. *Kennedy* (New York: Harper, 1965).

Strauss, Leo. *Thoughts on Machiavelli* (Chicago, IL: University of Chicago Press, 1958).

Sutphen, Mona, and Nina Hachigian. *The Next American Century: How the U.S. Can Thrive as Other Powers Rise* (New York: Simon and Schuster, 2008).

Talbott, Strobe. *The Great Experiment: The Story of Ancient Empires, Modern States, and the Quest for a Global Nation* (New York: Simon and Schuster, 2008).

Vance, Cyrus R. *Hard Choices: Critical Years in America's Foreign Policy* (New York: Simon and Schuster, 1983).

Zakaria, Fareed. *The Post-American World* (New York: Norton, 2008).

Articles

Bergsten, C. Fred. "Foreign Economic Policy for the Next President," *Foreign Affairs* (March/April 2004).

———. "A Partnership of Equals: How Washington Should Respond to China's Economic Challenge," *Foreign Affairs* (July/August 2008).

Buruma, Ian. "After America: Is the West Being Overtaken by the Rest?" *The New Yorker* (April 21, 2008).

Daalder, Ivo, and Robert Kagan. "The Next Intervention," *The Washington Post* (August 6, 2007), A17.

Dahl, Robert. "The Concept of Power," *Behavioral Science*, 2:3 (July 1957).

Haass, Richard N. "The Age of Nonpolarity: What Will Follow U.S. Dominance," *Foreign Affairs* (May/June 2008).

Hoffmann, Stanley. "International Organization and the International System," *International Organization*, 24:3 (Summer 1970).

Ikenberry, G. John. "America's Imperial Ambition," *Foreign Affairs* (September/October 2002).

———. "Illusions of Empire: Defining the New American Order," *Foreign Affairs* (March/April 2004).

Kagan, Robert. "The End of the End of History: Why the Twenty-First Century Will Look Like the Nineteenth," *The New Republic* (April 23, 2008).

———. "The September 12 Paradigm: America, the World, and George W. Bush," *Foreign Affairs* (September/October 2008).

Krauthammer, Charles. "The Unipolar Moment Revisited," *The National Interest* (Winter 2002).

Scowcroft, Brent. "The Dispensable Nation?" *International Herald Tribune* (August 7, 2007).

Walt, Stephen M. "Taming American Power," *Foreign Affairs* (September/October 2005).

X. "The Sources of Soviet Conduct," *Foreign Affairs* (July 1947).

Zakaria, Fareed. "The Future of American Power: How America Can Survive the Rise of the Rest," *Foreign Affairs* (May/June 2008).

Reports

Armitage, Richard, and Joseph S. Nye, Jr. "CSIS Commission on Smart Power: A Smarter, More Secure America," Center for Strategic and International Studies (November 6, 2007).

Ikenberry, G. John, and Anne-Marie Slaughter. "Forging a World of Liberty under Law: U.S. National Security in the 21st Century," Final Report of the Princeton Project on National Security (September 27, 2006).

Office of the Director of National Intelligence. "National Intelligence Estimate: Prospects for Iraq's Stability: A Challenging Road Ahead" (January 2007).

Speeches and Interviews

Kagan, Robert. "Why Should Democracy Be Shy?" interview, *Newsweek* (June 9, 2008).

Krauthammer, Charles. "Democratic Realism: An American Foreign Policy for a Unipolar World," speech at the American Enterprise Institute (February 10, 2004).

INDEX

Abdullah II, king of Jordan, 259
Abu Ghraib, 223, 228
Acheson, Dean, 44, 72, 296
  and containment policy, 47–48, 51
  Kennan-Acheson debates, 45,
    47–51, 57, 68
  and NSC-68, 50
  and postwar Soviet Union, 47–51,
    57
  and Vietnam War, 50
Afghanistan:
  al-Qaeda in, 66, 109, 170, 180, 290
  economic aid to, 180, 209, 210
  exit strategy for, 252
  insurgencies in, 87, 170
  and Iran, 247–48, 266
  and military power, 168–70
  nation-building attempts in, xv–xvi,
    23, 35, 165, 175
  NATO in, 19, 170, 179
  and Pakistan, 79, 181–82
  power package applied to, 179–80
  and regime change, 245
  and Saudi Arabia, 78
  Soviet invasion of, 15, 34, 58, 60, 78,
    106–7, 124, 170, 288
  stalemate in, 162
  Taliban in, 17–18, 19, 34, 66, 87,
    109, 165, 170, 180, 181–82, 248,
    268, 269
  U.S. forces in, 66, 79, 109, 165–66,
    170, 175–76, 290
  U.S. military aid to, 59
  U.S. strategy for, 290–91
Africa:
  AIDS in, 109
  U.S. military aid to, 209
  wars of liberation in, 247
Ahmadinejad, Mahmoud, 93, 94, 95,
  129, 251, 265
Albright, Madeleine, 113–14, 204
Alexander the Great, 33
Alliance for Progress, 210
al-Qaeda:
  in Afghanistan, 66, 109, 170, 180,
    290
  and power package, 180
  and September 11 attacks, 17, 34,
    66, 109, 290
  and Taliban, 17–18, 66, 170, 180
  and war on terror, 34, 109, 165, 170,
    224
Alsop, Stewart, 144
American Enterprise Institute, 142,
  238
American Revolution, 8–9
Americas:
  Alliance for Progress, 210
  nationalism in, 8
  U.S. interventions in, 14, 52, 60, 65,
    80, 172, 245
  wars of liberation in, 247
  see also specific nations
Ames, Aldrich, 137
Annan, Kofi, 21
Arafat, Yasir, 125, 274, 289

and Middle East, 102, 105–6, 210, 256, 257

and triangular diplomacy, 256

and Vietnam War, 14, 35, 55, 56, 58, 104–5, 106, 140, 153, 255–56, 287

and Watergate, 55

nongovernmental organizations (NGOs), 80–81, 143–44, 282

Non-State Actors, in power pyramid, 80–81

Noriega, Manuel, 175

Northern Ireland, 42, 65, 108, 260, 272–73

North Korea:

in axis of evil, 18, 67

bank accounts seized, 204

and China, 78, 87, 110, 200

economic aid to, 210

and Japan, 201

nuclear capabilities of, 18–19, 77–78, 87, 90, 101, 108, 109–10, 114, 118, 127–28, 162, 179, 200, 246, 268, 289

and regime change, 245

and South Korea, 79, 184

U.S. talks with, 65, 87, 179

U.S. threats against, 71, 163, 246, 289

Norway, in power pyramid, 79

Nuclear Nonproliferation Treaty (NPT), 268

nuclear power:

deterrence to use of, 4, 5, 36, 70–71, 90

neutron bomb, 57

power package vs., 179

threat of, 51, 67, 87, 90, 154, 240, 297

UN inspections, 65, 211, 222, 268

see also weapons of mass destruction

Nunn, Sam, 72, 299

Nye, Joseph, 23, 45, 68–71

Obama, Barack, 68, 126, 138, 156, 206

Official Development Assistance (ODA), 210

Oil and Gas Pumpers, in power pyramid, 78, 89

Olympic Games (2008), 111

Orwell, George, *Burmese Days*, 32

Ottoman Empire, 193

Overseas Private Investment Corporation (OPIC), 215

Pakistan:

and Afghanistan, 79, 181–82

economic aid to, 209

and military force, 171–72, 176

nuclear weapons in, 79, 82, 118, 162, 268, 269

in power pyramid, 79

as security threat, 181–82

Taliban in, 79, 170, 181–82, 268

and terrorism, 125

Palestine:

Arab supporters of, 78, 197

economic aid to, 39

and Israel, 150, 260, 272–75

and peace talks, 65, 125, 260, 272

territory of, 150

Panama, 75, 174–75

Panama Canal, 154, 249

Paulson, Henry, 204

Peloponnesian War, 11

Pentagon:

intelligence efforts of, 128

and Rwanda, 174, 289

war scenarios in, 67, 164–65

Peru, 210, 295

Philby, Kim, 137

Poland, 223

policy:

aim of, 231

and diplomacy, 231, 234, 242

political viability of, 284

related problems of, 247

and stage setting, 231–32, 233, 234

use of term, 231–32, 237

see also foreign policy

political economy, study of, 218

Portugal, empire of, 193

Powell, Colin L., 109, 222, 251